I0443721

Elements of Classical Point Mechanics
Volume I

Elements of Classical Point Mechanics

Volume I: Dirac's Theory of Constraints

Piotr W. Hebda

University of North Georgia, USA

ISBN 9798327510104 (Hardcover)
ISBN 9798326960832 (Paperback)

Elements of Classical Point Mechanics
Volume I: Dirac's Theory of Constraints

Copyright © 2024 Piotr W. Hebda, Beata A. Hebda

All rights reserved. This book or any part of it may not be reproduced in any form or by any means, mechanical or electronic, or in any retrieval system or information storage known or to be invented, without written permission from the copyright holders.

Dedicated to my mother, Halina Maria Hebda.

Acknowledgments

I would like to thank my wife, Beata, for reading and discussing the manuscript.

I would like to thank Professor Robert L. Anderson from Physics Department of the University of Georgia who, as my Ph.D. dissertation advisor, allowed me the freedom to work in the area of higher order Lagrangians, which started my lifelong fascination with Dirac's Theory of Constraints.

I would like to thank Professor Malcolm R. Adams from Mathematics Department of the University of Georgia, who in addition to being my master's thesis advisor, spent many hours discussing multiple topics essential for this book.

Elements of Classical Point Mechanics

Volume I: Dirac's Theory of Constraints

Contents:

Acknowledgements   v

INTRODUCTION

The Content of the Book.. 1
Description of Volume I... 3
Description of Volume II.. 4
Description of Volume III... 5

CHAPTER I. TIME, POSITIONS, VELOCITIES, AND THE NEWTONIAN EQUATIONS OF MOTION. THE ASSUMPTION OF REGULARITY

1. Time, Positions, Velocities... 7
2. The Newtonian Equations of Motion for Regular Mechanical Systems..................... 9

CHAPTER II. THE LAGRANGIAN AND THE EULER-LAGRANGE EQUATIONS OF MOTION

1. The Lagrangian and the Euler-Lagrange Equations of Motion................................. 14
2. An Example of a Lagrangian and its Euler-Lagrange Equations of Motion............. 16
3. Comments on Lagrangians and their Euler-Lagrange Equations of Motion............. 17

CHAPTER III. A HYPERREGULAR LAGRANGIAN, ITS HAMILTON'S EQUATIONS OF MOTION, ITS HAMILTONIAN, ITS POISSON BRACKETS, AND ITS NEWTONIAN EQUATIONS

1. The Legendre Transformation, the Hyperregular Case.................... 19
2. Obtaining the Hamilton's Equations of Motion from the Euler-Lagrange Equations, the Hyperregular Case 22
3. The Hamiltonian, the Hyperregular Case.. 24
4. Partial Derivatives of the Hamiltonian, the Hyperregular Case......... 25
5. Hamilton's Equations of Motion Obtained Directly from the Hamiltonian, the Hyperregular Case.. 28
6. The Poisson's Brackets, the Hyperregular Case................................. 29
7. The Newtonian Equations of Motion, the Hyperregular Case........... 31
8. Some Remarks.. 32

CHAPTER IV. AN EXAMPLE OF CONSTRAINED LAGRANGIAN, THE HAMILTONIAN SYSTEM OBTAINED FROM IT, THE COMPLETION PROCESS, AND THE DIRAC'S BRACKETS

1. An Example of a Constrained Lagrangian, its Euler-Lagrange Equations, the Constraints, and the Completion Process....................................... 33
2. An Example of a Constrained Lagrangian, the Legendre Transformation............. 36
3. An Example of a Constrained Lagrangian, the Hamilton's Equations of Motion..... 37
4. An Example of a Constrained Lagrangian, the Hamiltonian.............. 43
5. An Example of a Constrained Lagrangian, the Poisson Brackets and the Hamilton's Equations of Motion.. 49
6. An Example of a Constrained Lagrangian, the Newtonian Equations of Motion...... 50
7. An Example of a Constrained Lagrangian, the Dirac's Brackets........ 51
8. An Example of a Constrained Lagrangian, the Hamiltonian H_1 and the Dirac's Brackets... 55

CHAPTER V. FROM THE EULER-LAGRANGE EQUATIONS TO THE HAMILTON'S EQUATIONS AND THE NEWTONIAN EQUATIONS FOR REGULAR CONSTRAINED LAGRANGIANS

1. Regular Constrained Lagrangian, the Hamilton's Equations................................. 59
2. Regular Constrained Lagrangian, the Newtonian Equations................................. 67
3. A Comment.. 71

CHAPTER VI. GENERAL DESCRIPTION OF CONSTRAINED FUNCTIONS. EQUATIONS OF MOTION FOR CONSTRAINED FUNCTIONS

1. An Example of Functions Equal on Constraints... 72
2. Generalized Substitutions for a Single Constraint.. 74
3. Generalized Substitutions for Multiple Constraints...................................... 76
4. An Example of Derivatives for Functions Equal on Constraints................... 79
5. Derivatives of Functions Equal on Constraints.. 82
6. Effects of Constraints on the Equations of Motion...................................... 83

CHAPTER VII. POISSON BRACKETS FOR CONSTRAINED SYSTEMS. EFFECTS OF CONSTRAINTS ON THE HAMILTON'S EQUATIONS EXPRESSED BY POISSON BRACKETS

1. Definition of the Poisson Brackets for Systems with Constraints..................... 91
2. Calculations of Basic Properties of Poisson Brackets for Systems with Constraints.. 91
3. The Poisson Brackets Expressed in Different Sets of Coordinates................... 98
4. List of Basic Properties of Poisson Brackets for Systems with Constraints........ 104
5. Poisson Brackets for Functions Equal on Constraints................................. 105
6. Hamilton's Equations Using the Poisson Brackets for Functions Equal on Constraints.. 107

CHAPTER VIII. RETRIVING A FUNCTION FROM ITS PARTIAL DERIVATIVES

1. Retrieving a Function from its Partial Derivatives..................................... 112
2. An Example... 114
3. A Comment... 117

CHAPTER IX. THE PRE-HAMILTONIAN FOR A CONSTRAINED LAGRANGIAN

1. Introduction of the pre-Hamiltonian for a Constrained Lagrangian.................. 118
2. A Simple Example.. 120
3. The pre-Hamiltonian as a Function of the Positions and Momenta Only............. 122
4. Some Remarks... 128

CHAPTER X. ELEMENTS OF LINEAR ALGEBRA

1. An Important Theorem.. 130

CHAPTER XI. THE HAMILTONIAN FOR A REGULAR CONSTRAINED LAGRANGIAN

1. The Differential of the pre-Hamiltonian and the Differentials of Legendre Transformations... 134
2. The Hamilton's Equations for a Constrained Lagrangian................................... 143
3. The Definition of the Hamiltonian... 147
4. The relation between the pre-Hamiltonian and the Hamiltonian......................... 149
5. Including the Secondary Constraints in the Hamiltonian................................... 150

CHAPTER XII. A SUMMARY – GETTING THE HAMILTONIAN FORMALISM FROM A REGULAR CONSTRAINED LAGRANGIAN

1. Getting the Hamiltonian Formalism from a Regular Constrained Lagrangian...,... 151

CHAPTER XIII. THE DIRAC'S BRACKETS

1. The Need for Modifications of the Poisson Brackets... 157
2. Separating the Variables into Two Classes.. 159
3. Constructing a Minimal Complete Set of Class II Constraints........................... 160
4. Definition of the Dirac's Brackets... 167
5. Basic Properties of the Dirac's Brackets.. 168
6. Special Properties of the Dirac's Brackets... 169
7. The Dirac's Brackets Expressed in Different Variables...................................... 178
8. Obtaining the Newtonian Equations Using the Dirac's Brackets........................ 179

CHAPTER XIV. AN INTERPRETATION OF THE DIRAC'S BRACKETS

1. An Interpretation of the Dirac's Brackets.. 181

CHAPTER XV. COMMENTS ON DIRAC'S BRACKETS

1. Calculating the Dirac's Brackets One Pair or Few Pairs of Constraints at a Time..193
2. Using Only Some Constraints to Define the Dirac's Brackets.......................... 194
3. Imposing Additional Constraints on The System.. 194
4. Imposing Constants of Motion as Constraints... 195
5. A Comment about Volumes II and III.. 197

APPENDICES:

A. THE FLOW BOX THEOREM.. 198
B. THE DOUBLE FLOW BOX THEOREM... 206
C. THE DARBOUX THEOREM, THE VERSION FOR THE POISSON BRACKETS... 213
D. THE DARBOUX THEOREM, THE VERSION WITH CONSTRAINTS, AS APPLIED TO DIRAC'S BRACKETS... 224
E. THE DARBOUX-BOX VARIABLES.. 226
F. CALCULATIONAL SHORTCUTS FOR DIRAC'S BRACKETS........................... 228
G. AN INDEPENDENT DEFINITION OF THE POISSON BRACKETS................. 231
H. THE LEGENDRE TRANSFORMATIONS IN NON-TYPICAL VARIABLES.... 234

SUGGESTED READING.. 239

ABOUT THE AUTHOR 240

OUTLINE OF THE CONTENT OF VOLUME II

CHAPTER XVI. INDEPENDENCE OF THE LAGRANGE – HAMILTONIAN-DIRAC'S FORMALISM FROM THE CHOICE OF THE SPATIAL COORDINATES

CHAPTER XVII. A LAGRANGIAN THAT INCLUDES TWO OR MORE SETS OF SPATIAL COORDINATES (CHANGING COORDINATES VIA LAGRANGE MULTIPLIERS)

CHAPTER XVIII. TIME INDEPENDENT EXTERNAL CONSTRAINTS, GENERAL DESCRIPTION

CHAPTER XIX. IMPOSING EXTERNAL TIME INDEPENDENT CONSTRAINTS VIA LAGRANGE MULTIPLIERS

CHAPTER XX. IDEAL TIME INDEPENDENT EXTERNAL HOLONOMIC AND NONHOLONOMIC CONSTRAINTS

CHAPTER XXI. EXTERNAL TIME INDEPENDENT HOLONOMIC CONSTRAINTS IMPOSED VIA LAGRANGE MULTIPLIERS, THE DIRAC'S FORMALISM

CHAPTER XXII. REMOVING VELOCITIES FROM THE MAIN BODY OF A REGULAR, POSSIBLY CONSTRAINED, LAGRANGIAN

CHAPTER XXIII. HIGHER ORDER LAGRANGIANS

OUTLINE OF THE CONTENT OF VOLUME III

CHAPTER XXIV. AN EXAMPLE OF THE CHANGE OF VARIABLES IN THE LAGRANGIAN AND HAMILTONIAN FORMALISMS

CHAPTER XXV. DIRECT CHANGE OF VARIABLES IN THE HAMILTONIAN FORMALISM

CHAPTER XXVI. USING LAGRANGE MULTIPLIERS TO CHANGE VARIABLES

CHAPTER XXVII. SPECIAL SETS OF VARIABLES

CHAPTER XXVIII. EXPLICIT CONSTRUCTION OF THE DARBOUX-BOX VARIABLES AND THE DARBOUX VARIABLES

CHAPTER XXIX. CONDITIONS IMPOSED ON POISSON BRACKETS BY GIVEN SET OF EQUATIONS OF MOTION, EXPLICIT CONSTRUCTION OF THE HAMILTONIAN

CHAPTER XXX. ANY CONSTANT OF MOTION IS A HAMILTONIAN WITH PROPERLY CHOSEN SET OF POISSON BRACKETS

CHAPTER XXXI. THE STRUCTURE-IMPOSING LAGRANGIANS, THEIR LAGRANGE-HAMILTON FORMALISM, AND THEIR EXPLICIT FORM OF THE DARBOUX-FREE VARIABLES

CHAPTER XXXII. EXISTENCE AND CONSTRUCTION OF A LAGRANGIAN FOR A GIVEN SYSTEM OF EQUATIONS OF MOTION

CHAPTER XXXIII. EXISTENCE AND CONSTRUCTION OF A LAGRANGIAN FOR A SYSTEM WITH GIVEN POISSON BRACKETS AND A HAMILTONIAN

CHAPTER XXXIV. REPLACING AN EXISTING LAGRANGIAN WITH A STRUCTURE-IMPOSING LAGRANGIAN

CHAPTER XXXV. THE FUNDAMENTAL THEOREM OF CLASSICAL MECHANICS- ISOMORPHISM OF ANY TWO MECHANICAL SYSTEMS OF THE SAME DIMENSION

CHAPTER XXXVI. IMPOSING ISOMORPHISMS OF MECHANICAL SYSTEMS VIA COVERING LAGRANGIANS

CHAPTER XXXVII. AN EXAMPLE – A LAGRANGIAN FOR A NONHOLONOMIC SYSTEM, THE ADJUSTABLE CHAPLYGIN SLEIGH

CHAPTER XXXVIII. AN EXAMPLE – A COVERING LAGRANGIAN FOR A FREE PARTICLE, A HARMONIC OSCILLATOR AND A FREE-FALLING MASS

CHAPTER XXXIX. A LAGRANGIAN FOR THE ENTIRE UNIVERSE

INTRODUCTION

This book requires basic knowledge of multivariable calculus, but other than that, it is essentially self-contained, so even people without much knowledge of Classical Mechanics should be able to understand it. Still, motivation and determination are needed to go through the large volume of calculations we present.

We believe that a more-detailed, even if longer, mathematical explanation may be easier to read than a shorter text that leaves many details to be worked out by the reader. We are of the opinion that leaving details aside may get readers "stuck" on otherwise trivial details and waste a lot of their time. We try to avoid this, and the cost is a longer text.

We were able to present these details only because we used the "copy, paste, and modify" method to create the calculations shown in this book. We copied the current step, pasted it to the next line, and then created a new step by modifying what was pasted. Such a method helps to avoid simple copying errors. It also helps an interested reader see all the details of the material presented. It may also annoy someone who wants to go through the material faster. To help the latter, we have included a word explanation of steps, so it's easier to understand what is happening when just skimming over the text.

We tried to keep each chapter to be readable on its own as much as possible. So, if we need a specific formula or statement from a previous chapter, we direct the reader specifically to that statement or formula, to help save the reader's time and effort.

The Content of the Book

In the classical mechanics of material points we study how material points change positions in space and time.

We do not attempt to define the meaning of "time" in a general, physical sense. In our text, time, denoted by t, is a one-dimensional, real-number parameter used to organize the description of changes in positions of material points. Different parameters could possibly be used as "time," and we will not try to define the best one, or a "physical" one.

If a reader insists on a physical interpretation, perhaps the best will be the proper time associated with a single, permanently chosen observer. However, we will not attach too much meaning to this concept, since this book is not going to be detailed enough to specify if we consider the so called non-relativistic (Newtonian) systems or relativistic (Einsteinian) systems. Our approach will cover either, or possibly even systems with different kinds of relativity.

We would not attempt to define the meaning of the "material point." We simply recall that one, possibly the simplest, model to describe the universe is to assume that universe is made up of a finite number of impossible to create or destroy points which we call the "material points." These points have zero size, so they are also points in a mathematical sense. The points are moving in a fixed space, usually assumed to be three-dimensional.

It is typically assumed that all material points are moving in the same single common space. This assumption reflects our basic observation of the world around us. Mathematically, however, it is irrelevant if this is the same space, or if we have a separate space for each point. From the point of view presented in this book, it is not even important if all the spaces have the same dimensions.

We pay no attention to the specifics described above, and give positions of the entire system as a single point in N - dimensions (N referring to the space dimension) using generalized coordinates $(q_1, q_2, ..., q_N)$. We will assume that N is finite, and this is what we mean by studying a system of mechanical points. In general, we will not assume any specific decomposition of the variables $(q_1, q_2, ..., q_N)$ into separate "physical" particles. For example, if $N = 6$, we could separate $(q_1, q_2, ..., q_6)$ into $(q_1, ..., q_3)$ and $(q_4, ..., q_6)$, and interpret each triple as describing a separate three-dimensional "physical" particle. We are not interested in looking at the details of such possible decompositions; we usually just consider the entire $(q_1, q_2, ..., q_N)$ as one mechanical object. When describing such a system in different variables, we in general freely mix variables associated with "separate material particles," without paying any attention to the identity of each particle.

Only in specific applications, when we want to relate to "real" physics, we may be starting with variables being separated and associated with separate material particles. However, even in such cases, we will tend to mix all variables together at some stage of analysis.

In Volume III of this book, the reader will see that freedom of changes of variables is essential for our approach to mechanical systems and goes much further than just mixing coordinates of different "physical" particles. In this text, we are going toward the situation where we even stop distinguishing between the positions and velocities of mechanical systems, mixing them all together. Such an approach may be somewhat radical, but because it offers, in our opinion, a much deeper understanding of mechanical systems, we will insist on using it.

The classical mechanics of point particles is such a large subject that we were only able to cover a small part of it in this book. We concentrate on three major topics here. In Volume I, it is Dirac's Theory of Constrained Systems. In Volume II, it is inclusion of Lagrange multipliers as independent variables of mechanical systems. In Volume III, the topic is generalized variables and the role of an observer.

The reader will notice that we are not using the modern language of manifolds, tangent and cotangent spaces, and so on. This is because, in our opinion, this language will not be helping with the topics considered here. The main reason is that we do not want to be restricted to a well-defined configuration space. The essence of what we present is an arbitrariness of what is considered to be positions and what is considered to be velocities. This is associated with our concept of an observer being an entity that collects information about the system, without automatically separating this information into categories, like positions, velocities, or momenta.

Description of Volume I

In Volume I, we concentrate on Dirac's theory of constrained systems. In his Yeshiva Lectures (*see the Suggested Reading section at the end of this volume*), Paul A. M. Dirac presented a description of obtaining a Hamiltonian system starting with a constrained Lagrangian. Dirac's description is genius, but also extremely brief. In our work, we cover essentially the same material that he did, but we attempt to fill in the gaps left by him. We hope this approach may explain Dirac's ideas to a larger audience. We attempt to show all steps, providing as many details as possible.

One significant detail that differentiates our approach from Dirac's is that we decided not to use the concept and symbols for "weak" equality, used extensively by Dirac. The "weak" equality means that two expressions are equal only for the points that satisfy constraints. We decided that such a description is not precise enough. It is not precise, because at different stages of developing the Hamilton's formalism from the Lagrangian, we keep adding new constraints at multiple steps.

Therefore, a problem arises which of these constraints are already utilized or not when we say or denote "weak" during these calculations. So, we decided that using a word "weak," with its associated symbol $\approx$, was simply not convenient, since each time we were to use it, we would have to include a note explaining which constraints were used for that particular occurrence. Instead, we will use the typical symbol $=$, including a description of it in words as needed.

Notice that even in basic mathematics, we do not use different symbols for "strong" and "weak" equalities, since we use the same symbol for "strong" equality, as in

$x + x = 2x,$

which is "strong" in the sense that it is true for all $x's$, and

$3x = 6,$

which is "weak" in the sense that it is true only for $x = 3$.

So, in our work, we will follow the usual approach of using the $=$ in the wider sense.

Also, in our work, we place more emphasis on the Euler-Lagrange equations compared to what was done by Dirac.

Description of Volume II

In Volume II, we concentrate on Lagrange multipliers. We treat Lagrange multipliers as independent position variables, equal in status to other position variables. Then we show how this approach leads to automatic removal of Lagrange multipliers and their respective velocities and canonical momenta by the constraints that arise from the modified Lagrangian. We also include some topics that we consider interesting, like changing coordinates using Lagrangian multipliers, a discussion of external constraints imposed on a given system, removing velocities from the main body of a given Lagrangian, and higher order Lagrangians.

Description of Volume III

In Volume III, we mostly concentrate on details of general changes of variables (not just coordinates) in a mechanical system. We introduce some special variable systems that may simplify analysis of mechanical systems. We also show necessary and sufficient conditions for given Poisson Brackets to be allowed for a given system of equations of motion. In addition, we show that any nontrivial constant of motion is a Hamiltonian, with properly chosen Poisson Brackets.

Still in Volume III, we introduce a concept of structure-imposing Lagrangians and study their Hamiltonian structure. We also show a (non-explicit) construction of a Lagrangian for a given system of equations of motion, and we show a (non-explicit) construction of a Lagrangian for a system with given Poisson Brackets and a Hamiltonian.

Moreover, in Volume III, we formulate the Fundamental Theorem of Classical Mechanics. This theorem shows that any two mechanical systems of the same dimension are locally isomorphic. This leads to the concept of the Multiverse, which is an idea that all possible Universes, at all possible times, are happening simultaneously, and are different representations, in different coordinates, of what we observe as just one Universe: our Universe. Obviously, this is done only at the level of Classical Mechanics of points, which is a very imperfect model for the Universe. Still, many other models have Classical Mechanics at their base.

Another concept introduced in Volume III is the Covering Lagrangians. We present it there with some examples. Also in Volume III, we present the Lagrangian that is supposed to describe the motion of the entire Universe. Obviously, such Lagrangian would be the Holy Grail of Classical Mechanics. However, as may be expected, this Lagrangian, while possibly interesting from the theoretical point of view, seems to be rather useless in any practical application.

Finally, we would like to stress that all considerations in his book are "local," in the sense that they are presented using local systems of variables (in a manifold approach they would be associated with a singular local "chart" and its extension to tangent and cotangent spaces). So, we should point out that the word "local" we use is "mathematical," not "physical." It means that even the most distant galaxies are still "local" in the sense we use here. In this text, "local" refers to small changes around the given state of the mechanical system, not to the smallness of distances between the material points that make the mechanical system.

No artificial intelligence was used in the process of creating this book.

CHAPTER I

TIME, POSITIONS, VELOCITIES, AND THE NEWTONIAN EQUATIONS OF MOTION. THE ASSUMPTION OF REGULARITY

1. Time, Positions, Velocities

To describe positions for N - dimensional systems (N referring to a space dimension), we will use generalized coordinates $(q_1, q_2, ..., q_N)$. We will assume that N is finite. The coordinates $(q_1, q_2, ..., q_N)$ include all coordinates of the entire system so, for example, two material points, each moving in a three-dimensional space, will be described by $(q_1, q_2, ..., q_6)$.

To shorten the notation, instead of $(q_1, q_2, ..., q_N)$ we may use (q_i), or just (q). When we use the Cartesian coordinates, we may switch to (x_i) or (x_k, y_k, z_k). When working with specific low-dimensional examples, we may skip the subscript, and for example, use (x, y) for Cartesian coordinates, or (r, φ) for polar coordinates.

We assume that a mechanical system is in a specific position at any moment of time t, and that this position may change with time. For simplicity of considerations, we assume that this change is smooth, so that infinitely many time derivatives exist for allowed motions of the mechanical system. However, in specific applications, this requirement may be lowered down to only a small number of existing time derivatives, depending on what aspects of the system we want to study.

To denote velocities, by which we mean the derivatives of positions of a mechanical system with respect to the time parameter t, we will use generalized variables $(v_1, v_2, ..., v_N)$. To shorten the notation, we may use (v_i) or just (v). If convenient, we may also use symbols like (v_{x_i}). We will not have special symbols that will distinguish velocities calculated in Cartesian coordinates from velocities calculated using other coordinates. If we use (x, y) for position coordinates, we may use (v_x, v_y) for velocities, or we may use other symbols, for example (v, w).

In many texts, authors follow a convention according to which, when changing to new position coordinates, they automatically change the original velocity variables to the velocities (time derivatives) associated with the new position coordinates. We will not do this automatically here. We believe that it may be convenient to keep "old" velocity variables, even while using "new" position coordinates. The reader will have to follow each specific case.

Also, we will use various variable systems without trying to keep variables describing positions separated from variables describing velocities. For example, in a one-dimensional case, we may start with position-velocity variables (x, v) and then change to new variables (z_1, z_2), defined, for example, by:

$$z_1 = x + v,$$
$$z_2 = x - v. \tag{1.1}$$

The change of variables (1.1) is invertible:

$$x = \frac{1}{2}(z_1 + z_2),$$
$$v = \frac{1}{2}(z_1 - z_2). \tag{1.2}$$

We consider (z_1, z_2) to be legitimate variables for this example. Whether we use them will be based solely on how convenient they may be for a specific case under consideration.

This kind of change of variables, mixing the variables that describe velocities with variables that describe positions, may look strange or even unacceptable. However, such a mixing of variables is essential for what we want to present in this text. Also, such a mixing is less unusual than it may look since, as we are going to show in the later chapters, we routinely change from position-velocities systems of variables to position-momenta systems of variables, and momenta often mix positions and velocities.

Some readers may also feel uneasy about the fact that we combine variables that are measured using different physical units. For example, in (1.1), we add (or subtract) quantities possibly measured in meters to quantities possibly measured in meters per second. However, this is not a serious problem since each variable can always be multiplied by a separate constant, a constant with a numerical variable equal to 1. The constant may contain such physical units that after multiplication, the variable would have no physical units at all while having an unchanged numerical value. So, to simplify our considerations, we will always assume that all variables are unitless.

This approach is quite common. Even in trigonometry we typically change from degrees to radians when measuring angles, where the word "radian" means that there is no unit at all when describing angles. It then simplifies adding different powers of angles when using the Taylor series expansions for trigonometric functions. Moreover, in relativistic classical mechanics, we mix spatial-like and time-like variables, often using the speed of light to switch to unitless variables.

Finally, a remark about the wording – we will strictly reserve the word "coordinates" for variables that describe the positions of the system. All velocities, or combinations of positions and velocities, or just positions, will be described by the word "variables." Therefore, the set of "variables" includes the set of "coordinates."

2. The Newtonian Equations of Motion for Regular Mechanical Systems

Isaac Newton made this genius observation that to study possible motions of physical systems, we should look at solutions of ordinary differential equations, rather than directly at all possible motions. In this approach, a given physical system is characterized by a system of ordinary differential equations in which the time parameter t is the independent variable, and all positions and velocities play the role of dependent variables. Then, all possible motions of a physical system are given as all solutions of these differential equations.

The system of ordinary differential equations defining a physical system will be called the equations of motion for that system. We will assume that after a simplification process, which we will call the completion process, a subset of the equations will be solved for time derivatives of all velocities. So, we will always assume that after the completion process the system of equations will look like the following:

$$\frac{dv_1}{dt} = F_1(q_1, q_2, ..., q_N, v_1, v_2, ..., v_N),$$

$$\frac{dv_2}{dt} = F_2(q_1, q_2, ..., q_N, v_1, v_2, ..., v_N),$$

$$\vdots$$

$$\frac{dv_N}{dt} = F_N(q_1, q_2, ..., q_N, v_1, v_2, ..., v_N),$$

$$\frac{dq_1}{dt} = v_1,$$

$$\frac{dq_2}{dt} = v_2,$$

$$\vdots$$

$$\frac{dq_N}{dt} = v_N,$$

$$\phi_1(q_1, q_2, ..., q_N, v_1, v_2, ..., v_N) = 0,$$

$$\phi_2(q_1, q_2, ..., q_N, v_1, v_2, ..., v_N) = 0,$$

$$\vdots \qquad\qquad\qquad\qquad\qquad\qquad\qquad\qquad\qquad\qquad (1.4)$$

$$\phi_M(q_1, q_2, ..., q_N, v_1, v_2, ..., v_N) = 0, \qquad M \leq 2N.$$

The conditions $\phi_k(q_1, q_2, ..., q_N, v_1, v_2, ..., v_N) = 0$, $k = 1, ..., M$ in (1.4) are called constraints. After the completion process, they will contain no time derivatives. It is possible that a given system includes no constraints.

For convenience, the equations (1.4) can be written in a more concise form:

$$\frac{dv_i}{dt} = F_i(q_j, v_j),$$

$$\frac{dq_i}{dt} = v_i, \qquad\qquad\qquad\qquad\qquad\qquad\qquad (1.5)$$

$$\phi_k(q_j, v_j) = 0, \qquad i, j = 1, ..., N, \quad k = 1, ..., M \leq 2N,$$

or even more concise:

$$\dot{v}_i = F_i(q_j, v_j),$$
$$\dot{q}_i = v_i, \qquad\qquad\qquad\qquad\qquad\qquad\qquad (1.6)$$
$$\phi_k(q_j, v_j) = 0, \qquad i, j = 1,\ldots, N, \quad k = 1,\ldots, M \leq 2N,$$

where the time derivative is denoted by a dot above a variable, the second time derivative by two dots, and so forth, which is the notation commonly used by physicists.

Notice that the time derivatives of all positions appear on the left side of the equations with no position missing, which is not a restriction of any kind. These equations are just definitions of velocities.

Also, notice that time derivatives of all velocities appear on the left side of our equations, and no velocity is missing. This condition is not trivial, and it restricts the systems that we will study in this text.

Systems of equations of motion (1.6) in which time derivatives of some velocities are missing even after the completion process would allow the time evolution of these velocities to be arbitrary. This arbitrariness would then extend via other equations to other variables. Systems that will allow this kind of arbitrariness are called systems with gauge. Studying systems with gauge is a significant topic of its own. In this text however, we will only look at systems with no gauge.

A system in which no gauge is present is called a regular system. In this text we will study only regular systems.

The regularity of systems is related to the completion process, which we mentioned earlier. It is possible that the system of equations we begin with may not contain time derivatives of all velocities, which would suggest the existence of the gauge. Yet, it is also possible that interaction of the constraints with other equations will produce the missing equations in the completion process.

For example, if we have a constraint $v_x - v_y = 0$ in the system, and the equation containing $\dot{v}_y$ is missing from the system, then the completion process will tell us that since $v_x - v_y = 0$, then $\dot{v}_y$

must be equal to $\dot{v}_x$. This way, if the equation for $\dot{v}_x$ is given in the system, then the equation for $\dot{v}_y$ also becomes known and may be added to the original system. The system, in this way, may be completed, becoming a regular system.

Therefore, the decision that the system is not regular, and consequently should be discarded from our considerations, can only be made after the completion process is done, not before. The completion process will be described in more detail in Chapter IV and Chapter V.

Notice also that we assume that all time derivatives in the equations (1.6) are of the first order, and our equations do not contain derivatives of order two or higher. From the "mathematical" point of view it is not a significant restriction. We can always define new "positions" as equal to lower-order time derivatives of the "old" positions in the well-known procedure of lowering the order of systems of differential equations by introducing new variables. From the "physical" point of view, it is not that simple. Introducing new "positions" is contradictory to what we observe in the real world as "positions" of particles, which is a fundamental observation made by humans from the beginning of our species or even earlier.

Thus, if we decide to stick to the "physical" point of view, then in this text, we assume that the systems of differential equations of motion that we study are up to the second order (including, at most, the second derivative of positions with respect to time). We give up higher-order systems of equations. Luckily, it is not such a serious restriction since most systems of equations in Classical Mechanics are of the second order, and they become of the first order after we introduce velocities.

Systems like (1.6), are called Newtonian systems. The name is justified by the fact that the famous Newton's Second Law of Motion may be written as (we use 1-dimensional system for simplicity):

$$\ddot{x} = \frac{F}{m}, \qquad (1.7)$$

and can be rewritten as:

$$\dot{v} = \frac{F}{m}, \qquad (1.8)$$
$$\dot{x} = v.$$

Therefore, Newton's Second Law of Motion can be written in the form (1.6), with no constraints present. Hence, it is a Newtonian system as we defined it.

In Volume III, we will revisit the issue of what should be considered to be a "position" when we analyze a system of differential equations like (1.6), and we will see that the answer does not have to be trivial.

CHAPTER II

THE LAGRANGIAN AND THE EULER-LAGRANGE EQUATIONS OF MOTION

1. The Lagrangian and the Euler-Lagrange Equations of Motion

Assume that we have an N-dimensional mechanical system described by the variables $(q_1, q_2, ..., q_N, v_1, v_2, ..., v_N)$. One possible way to introduce equations of motion for the system is to define a function called the Lagrangian:

$$L = L(q_1, q_2, ..., q_N, v_1, v_2, ..., v_N), \tag{2.1}$$

together with the so-called Euler-Lagrange equations of motion defined as:

$$\frac{d}{dt}\left[\frac{\partial L}{\partial v_1}\right] = \frac{\partial L}{\partial q_1},$$

$$\frac{d}{dt}\left[\frac{\partial L}{\partial v_2}\right] = \frac{\partial L}{\partial q_2},$$

$$\vdots$$

$$\frac{d}{dt}\left[\frac{\partial L}{\partial v_N}\right] = \frac{\partial L}{\partial q_N},$$

$$\frac{dq_1}{dt} = v_1,$$

$$\frac{dq_2}{dt} = v_2,$$

$$\vdots$$

$$\frac{dq_N}{dt} = v_N. \tag{2.2}$$

It is convenient to write the Euler-Lagrange equations in a more concise form where, as before, the dot about a variable means the time derivative:

$$\frac{d}{dt}\left[\frac{\partial L}{\partial v_i}\right] = \frac{\partial L}{\partial q_i},\tag{2.3}$$
$$\dot{q}_i = v_i, \qquad i = 1,\ldots,N,$$

Throughout this text, we are not going to associate the Euler-Lagrange equations (2.1) with variational principles. We will not use variational principles at all in the remainder of this book. Instead, we will assume that obtaining the Euler-Lagrange equations (2.2) from a Lagrangian (2.1) is an axiom. In other words, we simply accept (2.2) as the equations of motion that are obtained from a Lagrangian (2.1) without any explanations. A reader interested in the variational aspect of obtaining the Euler-Lagrange equations from a given Lagrangian can easily find this topic in any introductory textbook on Classical Mechanics.

So, once we define a Lagrangian for a mechanical system, the resulting Euler-Lagrange equations become the equations of motion for that system and, therefore, also define all possible motions of that system.

The Euler-Lagrange equations (2.2) and (2.3) may look like complicated partial differential equations. This is not the case. We use these equations by starting with a given Lagrangian, calculating its partial derivatives first, and then we substituting these derivatives into the Euler-Lagrange equations. After the substitution, all partial derivatives are gone, and we deal with a set of ordinary differential equations.

A reader should be warned that modified forms of the equations (2.2) and (2.3) are being commonly studied, and they are often referred to as the Euler-Lagrange equations of motion. In this text, we will use the name "Euler-Lagrange equations" exclusively for the equations obtained from a given Lagrangian, using (2.2) or (2.3), with no modifications.

The equations (2.2) and (2.3) are not written in the Newtonian form. They can be, in principle, changed to the Newtonian form by the completion process, which will be described in more detail in Chapter IV and Chapter V.

2. An Example of a Lagrangian and its Euler-Lagrange Equations of Motion

Let us consider a two-dimensional mechanical system. We will describe its spatial variables by Cartesian coordinates (x, y), with their respective velocities given as (v, w). Assume that the Lagrangian is given by:

$$L(x, y, v, w) = \frac{v^2}{2} + \frac{w^2}{2} - \frac{x^2}{2} - \frac{y^2}{2}. \tag{2.4}$$

Since the partial derivatives of the Lagrangian (2.4) are:

$$\begin{aligned} \frac{\partial L}{\partial x} &= -x, \\ \frac{\partial L}{\partial y} &= -y, \\ \frac{\partial L}{\partial v} &= v, \\ \frac{\partial L}{\partial w} &= w, \end{aligned} \tag{2.5}$$

and the generic Euler-Lagrange equations of motion (2.3) in this case are:

$$\begin{aligned} \frac{d}{dt}\left[\frac{\partial L}{\partial v}\right] &= \frac{\partial L}{\partial x}, \\ \frac{d}{dt}\left[\frac{\partial L}{\partial w}\right] &= \frac{\partial L}{\partial y}, \\ \dot{x} &= v, \\ \dot{y} &= w, \end{aligned} \tag{2.6}$$

we get, as specific Euler-Lagrange equations:

$$\frac{d}{dt}[v] = -x,$$
$$\frac{d}{dt}[w] = -y,$$
$$\frac{dx}{dt} = v,$$
$$\frac{dy}{dt} = w.$$
(2.7)

Using the dot to denote the time derivative produces:

$$\dot{v} = -x,$$
$$\dot{w} = -y,$$
$$\dot{x} = v,$$
$$\dot{y} = w.$$
(2.8)

Notice that in this case the Euler-Lagrange equations (2.8) already in the Newtonian form (1.4), without the need to do any modifications. Since time derivatives of all variables are present in (2.8), the system contains no gauge. Also, there are no constraints.

3. Comments on Lagrangians and their Euler-Lagrange Equations of Motion

Without going into detail, let us mention that finding a Lagrangian for a given set of equations of motion may be very difficult. If we add more restrictions on how the Lagrangian must look like, there will exist sets of equations of motion that do not have any Lagrangian at all. Readers will find more on this subject in Volume II of this book.

On the other hand, it is also possible that a given system of equations of motion has more than one Lagrangian from which it can be obtained. Again, more on this subject will be found in Volume II.

Some Lagrangians may create the Euler-Lagrange equations that have no solutions. For example:

$$L(x,v) = v + x,$$
(2.9)

has partial derivatives $\frac{\partial L}{\partial x}=1$ and $\frac{\partial L}{\partial v}=1$. Substituting that into the Euler-Lagrange equations (2.3) gives:

$$\frac{d}{dt}[1]=1,$$
$$\dot{x}=v. \tag{2.10}$$

Since the time derivative of constant function is equal to zero, we get:

$$0=1,$$
$$\dot{x}=v. \tag{2.11}$$

The system of equations (2.11) is obviously never satisfied, so the Lagrangian (2.9) describes no motion of a system. Lagrangians like that will be removed from our considerations.

In this text, we will consider any Lagrangian that produces a given set of equations of motion by the equations (2.3) to be a legitimate Lagrangian for that set of equations (possibly not a unique one).

The reader should be aware that many researchers, especially physicists, impose additional, somewhat arbitrary restrictions on how a "correct" Lagrangian must look. Some researchers are so restrictive that they allow only Lagrangians that are created by subtracting the physical potential energy from the physical kinetic energy of a system. This may be a physically valid approach, but it restricts the use of Lagrangians to a quite narrow set of situations. Our approach in this text is on the other extreme - we allow any function to be a Lagrangian of a system, provided it creates correct equations of motion. Obviously, various points of view between these two extremes are possible.

Our approach of allowing all Lagrangians that give correct equations of motion should not be interpreted as the statement that they are all physically correct. It is rather the statement that, in our opinion, they are all worth looking at and studying.

© Hebda 2024

CHAPTER III

A HYPERREGULAR LAGRANGIAN, ITS HAMILTON'S EQUATIONS OF MOTION, ITS HAMILTONIAN, ITS POISSON BRACKETS, AND ITS NEWTONIAN EQUATIONS

In this chapter, we briefly review how the Hamilton's equations of motion, the Hamiltonian, and the Poisson Brackets are obtained from a non-constrained (also called hyperregular), Lagrangian. This material can be found in almost every essential book on Classical Mechanics, although typically with fewer details than we present here. We include this material here for completeness since the main topic of this book is to extend these constructions to the case of constrained Lagrangians.

1. The Legendre Transformation, the Hyperregular Case

In Chapters I and II, we describe mechanical systems using the position-velocity variables. In many situations, using a different system of variables may be convenient. In principle, any system of variables is allowed as long as the new variables are in a one-to-one relationship, at least locally, with the original position-velocity variables. In this section we will study so-called "canonical variables."

Assume that a system is initially described by the position-velocity variables (q_i, v_i), $i = 1,...,N$, and its equations of motion are obtained from a Lagrangian $L(q,v)$ via its Euler-Lagrange equations (2.3), given in Chapter II. Then a different set of variables exists, (q_i, p_i), $i = 1,...,N$, which is especially convenient. These variables are called "canonical variables."

They are defined by:

$$q_i = q_i,$$
$$p_i = \frac{\partial L}{\partial v_i}, \qquad i = 1,...,N. \tag{3.1}$$

In (3.1) the "new" variables are on the left side, and the "old" variables are on the right side. Strictly speaking, the new and old variables should not be denoted by the same symbols. However, since in the case of the variable change (3.1), the "old" spatial variables (q_i) are always numerically equal to the "new" spatial variables (q_i), people use the same symbols for both, and we will also do that in this book.

Notice that it may cause confusion, especially when we calculate the partial derivatives with respect to a spatial variable. A partial derivative depends not only on the variable with respect to which the derivative is taken, but also on variables that are kept constant during the process. So, for a given function f, the partial derivative $\dfrac{\partial f}{\partial q_i}$ may give different results, depending on which system of variables, the (q_i, v_i) or/and (q_i, p_i), we are using when describing the function f.

Once we introduce the variables (q_i, p_i), potentially either of the two variable systems can be used at any place, creating possible confusion. It is even possible to use both (q_i, v_i) and (q_i, p_i) in the same expression, and we actually do it in some situations (for example, in the formulas (3.3) and (3.4) below).

So, we need to be sure that in the future, when we calculate partial derivatives, either the context makes it obvious which system is used or we state it explicitly.

As a step toward removing this possible confusion, let's make a general agreement now that says that when calculating partial derivatives of the Lagrangian, by default we treat the Lagrangian as a function of (q_i, v_i). This means that when calculating the derivatives $\dfrac{\partial L}{\partial q_i}$ and/or $\dfrac{\partial L}{\partial v_i}$ we use the variables (q_i, v_i), unless we clearly specify otherwise.

So, in the formulas (3.1), we use the variables (q_i, v_i) when calculating the partial derivatives. It follows from the agreement we just made. It should also be obvious from the context that since (3.1) is the original definition of the variables (q_i, p_i), we could not have used them yet.

The definition (3.1) may look complicated since it involves partial derivatives. However, when used in specific calculations, it is not complicated at all, since once the partial derivatives are explicitly calculated, we get formulas similar to any typical change of variables, containing no derivatives.

The change of variables (3.1) has its name. It is called the Legendre Transformation.

For the remainder of this chapter, we will only consider Lagrangians for which the formulas (3.1) are invertible (at least locally). So, the formulas (3.1) will allow an inverse written as:

$$q_i = q_i,$$
$$v_i = v_i(q, p), \qquad i = 1, \ldots, N. \tag{3.2}$$

Not every Lagrangian will automatically assure the existence of the inverse of (3.1). For example:

$$L = L(x, v) = xv$$

will give as its Legendre Transformation:

$$x = x,$$
$$p = \frac{\partial L}{\partial v},$$

which produces:

$$x = x,$$
$$p = x.$$

Obviously, the above cannot be inverted; it is not possible to get the velocity v as a function of (x, p), using the last two equations.

If a given Lagrangian produces the Legendre Transformation, which is an invertible change of variables, then such Lagrangian is called "hyperregular" or "non-constrained." Hyperregular Lagrangians often appear in a description of important physical systems.

In this book though, starting in Chapter IV, we will concentrate on Lagrangians that are not hyperregular. Such Lagrangians are also called "constrained". Extending the notion of the Legendre Transformation to the constrained Lagrangians is the essence of Dirac's Theory of Constraints.

2. Obtaining the Hamilton's Equations of Motion from the Euler-Lagrange Equations, the Hyperregular Case

Assume we have a hyperregular Lagrangian, and the Euler-Lagrange equations of motion obtained from it. Now we want to define the so called Hamilton's equations of motion.

The definition is simple - the Hamilton's equations of motion are the Euler-Lagrange equations of motion in which the position-velocities variables (q_i, v_i) are expressed by the canonical variables (q_i, p_i), using the formulas (3.1) and (3.2).

To get the Hamilton's equations, we start with the Euler-Lagrange equations (2.3) from Chapter II, and we replace the velocities v_i at the bottom half of equations by the formula expressing v_i by the inverse $v_i = v_i(q, p)$ from (3.2). We get:

$$\frac{d}{dt}\left[\frac{\partial L}{\partial v_i}\right] = \frac{\partial L}{\partial q_i},$$
$$\dot{q}_i = v_i(q, p), \qquad i = 1, \ldots, N. \tag{3.3}$$

Then, we use the definition $p_i = \dfrac{\partial L}{\partial v_i}$ from (3.1) to replace the left sides of the top half of the equations (3.3), getting:

$$\dot{p}_i = \frac{\partial L}{\partial q_i},$$
$$\dot{q}_i = v_i(q, p), \qquad i = 1, \ldots, N. \tag{3.4}$$

Finally, we replace all velocities v_i by the inverse formula $v_i = v_i(q,p)$ from (3.2), getting the final form of the Hamilton's equations:

$$\dot{p}_i = \left.\frac{\partial L}{\partial q_i}\right|_{v=v(q,p)},$$

$$\dot{q}_i = v_i(q,p), \qquad i = 1,...,N. \tag{3.5}$$

Notice that, because of the replacements we did, the Hamilton's equations (3.5) do not contain velocities v_i anymore. Also, notice that the partial derivatives $\frac{\partial L}{\partial q_i}$ in (3.5) must be calculated using the variables (q_i, v_i), and only after the calculation we go to the canonical variables (q_i, p_i). This is because these partial derivatives come from the original Euler-Lagrange equations (1.2), where the variables (q_i, v_i) were used, and we cannot just replace them with variables (q_i, p_i) and expect the partial derivatives to remain unchanged.

It is important to realize that the Hamilton's equations (3.5) are the same equations as the Euler-Lagrange equations (2.3). They are the same equations, just expressed in different variables. They have the same solutions, just expressed in different variables.

Notice that we have not defined the Hamiltonian yet, but we could already obtain the Hamilton's equations. We will proceed to define the Hamiltonian now and show how it relates to the Hamilton's equations.

Also, notice that at some stages of the change from the Euler-Lagrange equations to the Hamilton's equations, we may use both (q_i, v_i) and (q_i, p_i), as we did in the formulas (3.3) and (3.4). In such situations, it may be difficult to decide if we should describe the equations of motion as the Euler-Lagrange or the Hamilton's. In the future, in such situations, we may use either description.

3. The Hamiltonian, the Hyperregular Case

Assume we have a hyperregular Lagrangian $L = L(q,v)$. Using the definition of the canonical variables (3.1) and its inverse (3.2), we define the Hamiltonian as:

$$H(q,p) = \sum_{i=1}^{N} p_i \cdot v_i(q,p) - L(q,v) \Big|_{v_i = v_i(q,p)}. \tag{3.6}$$

We can also write (3.6) as:

$$H(q,p) = \sum_{i=1}^{N} p_i \cdot v_i(q,p) - L(q,v(q,p)). \tag{3.7}$$

Notice that despite the fact that in the definition we started with the mix of the (q_i, v_i) and (q_i, p_i) variables, we then use the replacement $v_i = v_i(q,p)$ from (3.2) (which was written here as $v = v(q,p)$ for convenience). These replacements got rid of all the v_i variables, so we consider the Hamiltonian to be the function of (q_i, p_i) variables only.

In a sense, it does not matter what the variables in a function are. It is still the same function no matter what variables we use. However, it matters when we calculate the partial derivatives.

From now on we will always assume, by default, that when calculating the partial derivatives $\frac{\partial H}{\partial q_i}$ and/or $\frac{\partial H}{\partial p_i}$ of the Hamiltonian (3.6) or (3.7), we will use the variables (q_i, p_i), unless we clearly specify otherwise.

Notice that this agreement is different than the agreement about the Lagrangian L which, as we agreed earlier, is considered to be a function of (q_i, v_i). This means that when calculating the derivatives $\frac{\partial L}{\partial q_i}$ and/or $\frac{\partial L}{\partial v_i}$ we use the variables (q_i, v_i), unless we clearly specify otherwise.

4. Partial Derivatives of the Hamiltonian, the Hyperregular Case

Assume we a have system governed by a hyperregular Lagrangian $L = L(q,v)$. Recall that the Hamiltonian is defined as in (3.7) by:

$$H(q,p) = \sum_{i=1}^{N} p_i \cdot v_i(q,p) - L(q,v(q,p)) \tag{3.8}$$

In (3.8), the symbols $q, p,$ and v represent q_j, p_j and v_j, $j = 1,...N$.

Now we want to calculate the partial derivatives of the Hamiltonian H. We want to calculate these derivatives in variables (q,p). We will use the expression (3.8) for the Hamiltonian, and this expression includes the Lagrangian. So, we have this somewhat unusual situation that we calculate the partial derivatives of the Lagrangian using (q,p) not (q,v). The calculation utilizes a somewhat confusing way of using the chain rule. The possible confusion is the result of the fact that we decided to use the same symbol "q" for the "old" and the "new" variables in the change of variables given in (3.1) and (3.2). For readers who find the calculations (3.9) and/or (3.12) confusing, we suggest checking Appendix H. In Appendix H, we are showing the same calculations using more traditional symbols for the change of variables (3.1) and (3.2).

Taking the partial derivative of the Hamiltonian (3.8) with respect to q_n, $n = 1,...,N,$ when we treat the Hamiltonian as a function of (q_i, p_i), using the definition (3.1) and the chain rule, we get:

$$\frac{\partial H}{\partial q_n} = \frac{\partial}{\partial q_n}\left[\sum_{i=1}^{N} p_i \cdot v_i(q,p)\right] - \frac{\partial L(q,v(q,p))}{\partial q_n} =$$

(continued on the next page)

$$= \sum_{i=1}^{N} \frac{\partial p_i}{\partial q_n} \cdot v_i(q,p) + \sum_{i=1}^{N} p_i \cdot \frac{\partial v_i(q,p)}{\partial q_n} - \frac{\partial L(q,v(q,p))}{\partial q_n} =$$

$$= \sum_{i=1}^{N} 0 \cdot v_i(q,p) + \sum_{i=1}^{N} p_i \cdot \frac{\partial v_i(q,p)}{\partial q_n} - \frac{\partial L(q,v(q,p))}{\partial q_n} =$$

$$= \sum_{i=1}^{N} p_i \cdot \frac{\partial v_i(q,p)}{\partial q_n} - \frac{\partial L(q,v(q,p))}{\partial q_n} =$$

$$= \sum_{i=1}^{N} p_i \cdot \frac{\partial v_i(q,p)}{\partial q_n} - \frac{\partial L}{\partial q_n} - \sum_{i=1}^{N} \frac{\partial L}{\partial v_i} \cdot \frac{\partial v_i(q,p)}{\partial q_n} = \qquad (3.9)$$

$$= \sum_{i=1}^{N} p_i \cdot \frac{\partial v_i(q,p)}{\partial q_n} - \frac{\partial L}{\partial q_n} - \sum_{i=1}^{N} p_i \cdot \frac{\partial v_i(q,p)}{\partial q_n} =$$

$$= -\frac{\partial L}{\partial q_n} \qquad\qquad n = 1,\ldots, N.$$

When calculating the derivative (3.9) the symbols $\frac{\partial L}{\partial q_n}$ and $\frac{\partial L}{\partial v_n}$ are treating the Lagrangian L as a function of (q_i, v_i), as agreed before. However, the expression $\frac{\partial L(q,v(q,p))}{\partial q_n}$ is treating the Lagrangian as a function of (q_i, p_i). Again, if this is confusing, please refer to Appendix H.

The final conclusion of the calculation (3.9) is that:

$$\frac{\partial H}{\partial q_n} = -\frac{\partial L}{\partial q_n}, \qquad n = 1,\ldots, N. \qquad (3.10)$$

Let us stress again that in the expression (3.10), the partial derivative $\frac{\partial H}{\partial q_n}$ is calculated with the assumption that H is the function of (q_i, p_i), while the derivative $\frac{\partial L}{\partial q_n}$ in the same expression is calculated with the assumption that L is the function of (q_i, v_i). So, the two partial derivatives in the (3.10) represent different operations, despite of using the same symbol $\frac{\partial}{\partial q_n}$. As we explained earlier, this is the result of using the same symbol q_n in two different systems of coordinates, the (q_i, v_i) and the (q_i, p_i) systems. To be mathematically correct, we should be using different symbol for $q's$ in each variable system. However, since the $q's$ are always numerically equal in both systems, most literature in physics uses the same symbol, and we decided to do the same here. *Please see Appendix H for more mathematically correct calculations.*

We should also notice that in the actual calculation, both sides of (3.10) will look differently since the left side will be expressed in variables (q_i, p_i), and the right side will be expressed in variables (q_i, v_i). But if we change both sides to the same variables, they will turn out to be identical. If we decide to use the variables (q_i, p_i) on both sides, then (3.10) can be written as:

$$\frac{\partial H}{\partial q_n} = -\frac{\partial L}{\partial q_n}\bigg|_{v=v(q,p)}, \qquad n=1,...,N. \qquad (3.11)$$

The calculation of the derivative of the Hamiltonian (3.8) with respect to the canonical momenta p_n, $n=1,...,N,$ goes as follows:

$$\begin{aligned}
\frac{\partial H}{\partial p_n} &= \frac{\partial}{\partial p_n}\left[\sum_{i=1}^{N} p_i \cdot v_i(q,p)\right] - \frac{\partial L(q,v(q,p))}{\partial p_n} = \\
&= \sum_{i=1}^{N} \frac{\partial p_i}{\partial p_n} \cdot v_i(q,p) + \sum_{i=1}^{N} p_i \cdot \frac{\partial v_i(q,p)}{\partial p_n} - \frac{\partial L(q,v(q,p))}{\partial p_n} = \\
&= \sum_{i=1}^{N} \delta_{ni} \cdot v_i(q,p) + \sum_{i=1}^{N} p_i \cdot \frac{\partial v_i(q,p)}{\partial p_n} - \frac{\partial L(q,v(q,p))}{\partial p_n} = \\
&= v_n(q,p) + \sum_{i=1}^{N} p_i \cdot \frac{\partial v_i(q,p)}{\partial p_n} - \sum_{i=1}^{N} \frac{\partial L}{\partial v_i} \cdot \frac{\partial v_i(q,p)}{\partial p_n} = \\
&= v_n(q,p) + \sum_{i=1}^{N} p_i \cdot \frac{\partial v_i(q,p)}{\partial p_n} - \sum_{i=1}^{N} p_i \cdot \frac{\partial v_i(q,p)}{\partial p_n} = \\
&= v_n(q,p) \qquad n=1,...,N.
\end{aligned} \qquad (3.12)$$

As we agreed earlier, the partial derivative $\frac{\partial H}{\partial p_n}$ in (3.12) is defined using variables (q_i, p_i).

Again, if the somewhat unusual form of the chain rule in the calculation (3.12) is confusing to the reader, Appendix H may help.

In the calculation (3.12), we used the Kronecker delta symbol defined as:

$$\delta_{ni} = \begin{cases} 1 & n = i \\ 0 & n \neq i, \end{cases}$$

and the fact that $\frac{\partial L}{\partial v_i} = p_i$ as defined in (3.1).

Concluding, we get:

$$\frac{\partial H}{\partial p_n} = v_n(q, p), \qquad n = 1, \ldots, N. \tag{3.13}$$

Notice that when actually calculating the partial derivative in (3.13), we will not get the velocity v_n explicitly, but we will get an expression for v_n in variables (q_i, p_i), the same expression that we got in (3.2).

5. Hamilton's Equations of Motion Obtained Directly from the Hamiltonian, the Hyperregular Case

Earlier, we obtained the Hamilton's equations of motion (3.5) from the Lagrangian. If we compare the Hamilton's equations (3.5) with the partial derivatives of the Hamiltonian (3.11) and (3.13), we can see that:

$$\begin{aligned} \dot{p}_i &= -\frac{\partial H}{\partial q_i}, \\ \dot{q}_i &= \frac{\partial H}{\partial p_i}, \qquad i = 1, \ldots, N. \end{aligned} \tag{3.14}$$

Just like the equations (3.5), the equations (3.14) are called the Hamilton's equations of motion. Equations (3.14) do not contain the step of taking partial derivative of the Lagrangian using the variables (q,v), as did the case of the equations (3.5). Nevertheless, they are the same equations of motion.

Thus, we have the change from the description that uses the Lagrangian, namely the Euler-Lagrange equations (2.3) from Chapter II, and (q,v) as the variables, to the description that uses the Hamiltonian (3.8), the Hamilton's equations (3.14), and (q,p) as the variables. Despite the change, the equations of motion are in fact identical, except that they are expressed in different systems of variables. Therefore, the solutions of these equations are also identical, just expressed in different systems of variables. These are two completely equivalent systems.

6. The Poisson's Brackets, the Hyperregular Case

Assume we have a solution $q_i = q_i(t)$, $p_i = p_i(t)$, $i=1,...,N$ of the Hamilton's equations of motion (3.14). Assume also that we have a function $F = F(q_i, p_i)$, $i=1.,...,N$. Then, we can consider F to be the function of time, by substituting into it the solutions above, and getting:

$$F = F(t) = F(q_i(t), p_i(t)), \qquad i=1.,...,N. \tag{3.15}$$

Strict mathematicians would consider this process to be a creation of a new function, related to F, but still a separate function, for which we should use a different symbol, not F. They are essentially right, but we do not want to be as picky, and we will follow what most physicists do, and use the symbol F for both. We want to avoid having too many different symbols in our book, hopefully avoiding confusion.

Using the chain rule, we can calculate the time derivative of the function $F(t)$ as:

$$\frac{dF}{dt} = \sum_{i=1}^{N}\left(\frac{\partial F}{\partial q_i}\cdot\frac{dq_i}{dt} + \frac{\partial F}{\partial p_i}\cdot\frac{dp_i}{dt}\right). \tag{3.16}$$

Using the notation with a dot above a symbol for time derivative, we can write (3.16) as:

$$\dot{F} = \sum_{i=1}^{N}\left(\frac{\partial F}{\partial q_i}\cdot\dot{q}_i + \frac{\partial F}{\partial p_i}\cdot\dot{p}_i\right). \tag{3.17}$$

Now replacing $\dot{q}_i$ and $\dot{p}_i$ in (3.17) by the right sides of the Hamilton's equations of motion (3.14) we get:

$$\dot{F} = \sum_{i=1}^{N}\left(\frac{\partial F}{\partial q_i}\cdot\frac{\partial H}{\partial p_i} - \frac{\partial F}{\partial p_i}\cdot\frac{\partial H}{\partial q_i}\right). \tag{3.18}$$

At this point, it is convenient to introduce a new mathematical object, called the Poisson Bracket. The Poisson Bracket assigns, for an ordered pair of functions $F = F(q,p)$ and $G = G(q,p)$, a new function of (q,p) denoted by $\{F,G\}$.

The Poisson Bracket is defined as:

$$\{F,G\} = \sum_{i=1}^{N}\left(\frac{\partial F}{\partial q_i}\cdot\frac{\partial G}{\partial p_i} - \frac{\partial F}{\partial p_i}\cdot\frac{\partial G}{\partial q_i}\right). \tag{3.19}$$

Using the definition (3.19) of the Poisson Bracket for a function F and the Hamiltonian H we get

$$\{F,H\} = \sum_{i=1}^{N}\left(\frac{\partial F}{\partial q_i}\cdot\frac{\partial H}{\partial p_i} - \frac{\partial F}{\partial p_i}\cdot\frac{\partial H}{\partial q_i}\right). \tag{3.20}$$

Comparing that with the equation (3.18) gives us:

$$\dot{F} = \{F,H\}. \tag{3.21}$$

Since F can be any function of (q,p) then, in particular, we may have $F(q,p)=q_i$ or $F(q,p)=p_i$. Placing this in (3.21) gives us:

$$\dot{q}_i = \{q_i, H\},$$
$$\dot{p}_i = \{p_i, H\}. \tag{3.22}$$

The formula (3.22) is another form in which Hamilton's equations of motion (3.14) can be written.

7. The Newtonian Equations of Motion, the Hyperregular Case

The Newtonian Equations (1.4) are important to have, since they use the variables (q,v), which are directly related to physical observations. Sometimes, as the example in section 2 of Chapter II shows, the Euler-Lagrange equations lead to Newtonian equations directly. This is the case in many physically important examples. However, with more complicated Lagrangians it may not be so obvious that their Euler-Lagrange equations lead to the Newtonian equations, even in the case of the hyperregular Lagrangian. So, it may be worth discussing it.

From the definition of velocities, or by looking at the earlier Euler-Lagrange equations (2.3) in Chapter II, we have:

$$v_i = \dot{q}_i, \qquad i=1,\ldots,N. \tag{3.23}$$

Using the first line of (3.22) in (3.23), we get:

$$v_i = \{q_i, H\}. \tag{3.24}$$

Then, taking the time derivative of both sides of (3.24) and using (3.21), we get:

$$\dot{v}_i = \{\{q_i, H\}, H\}, \qquad i=1,\ldots,N. \tag{3.25}$$

The equations (3.25) are the Newtonian equations for the system given by the hyperregular (unconstrained) Lagrangian (3.1) or given by the Hamiltonian (3.8).

Using (3.23) to replace v_i by $\dot{q}_i$ in (3.25), we get the Newtonian equations written in the customary second-order form:

$$\ddot{q}_i = \{\{q_i, H\}, H\}, \qquad i = 1, \ldots, N. \tag{3.26}$$

8. Some Remarks

Observe that if we start with a given Hamiltonian, then (3.24) gives us the velocity expressed by positions and momenta. We also get the Newtonian equations of motion as (3.25) or (3.26).

Observe that the Newtonian equations obtained from a hyperregular Lagrangian contain no constraints.

Also, observe that the formulas (3.25) contain all velocities. This means that the hyperregular Lagrangian is automatically regular. In other words, a hyperregular Lagrangian has no gauge.

CHAPTER IV

AN EXAMPLE OF CONSTRAINED LAGRANGIAN, THE HAMILTONIAN SYSTEM OBTAINED FROM IT, THE COMPLETION PROCESS, AND THE DIRAC'S BRACKETS

1. An Example of a Constrained Lagrangian, its Euler-Lagrange Equations, the Constraints, and the Completion Process

In Chapter III, we described a mathematical structure we can build when starting with a, so-called, "hyperregular" Lagrangian. Here we want to show an example of how we adapt this structure to a Lagrangian that is not hyperregular. This example is intended to be possibly the simplest that will present typical difficulties when analyzing a constrained Lagrangian. In later chapters, we will give a general approach.

Assume we have a system described by two space variables (x, y), with velocities denoted respectively by (v, w). Assume also that the Lagrangian of this system is given as:

$$L = L(x, y, v, w) = xw - yv - x^2 y^2 \qquad (4.1)$$

With this Lagrangian, the Euler-Lagrange equations (2.3), as described in Chapter II, are:

$$\begin{aligned}
\frac{d}{dt}\left[\frac{\partial L}{\partial v}\right] &= \frac{\partial L}{\partial x}, \\
\frac{d}{dt}\left[\frac{\partial L}{\partial w}\right] &= \frac{\partial L}{\partial y}, \\
\frac{dx}{dt} &= v, \\
\frac{dy}{dt} &= w.
\end{aligned} \qquad (4.2)$$

Calculating the specific partial derivatives of the Lagrangian (4.1) and substituting the results into (4.2), we get:

$$\frac{d}{dt}[-y] = w - 2xy^2,$$
$$\frac{d}{dt}[x] = -v - 2x^2 y,$$
$$\frac{dx}{dt} = v,$$
$$\frac{dy}{dt} = w.$$
(4.3)

It is convenient to multiply the first equation above by (-1) and replace in all equations the symbol $\frac{d}{dt}$ with a dot above the expression that is being differentiated. We get:

$$\dot{y} = -w + 2xy^2,$$
$$\dot{x} = -v - 2x^2 y,$$
$$\dot{x} = v,$$
$$\dot{y} = w.$$
(4.4)

It is easy to see that the left sides of the first and last equation in (4.4) are equal. Therefore, their right sides must be equal as well. Similarly, the left sides of the second and third equations in (4.4) are equal, so their right sides must also be equal. So, we can add two more equations to the system (4.4), getting:

$$\dot{y} = -w + 2xy^2,$$
$$\dot{x} = -v - 2x^2 y,$$
$$\dot{x} = v,$$
$$\dot{y} = w,$$
$$w = -w + 2xy^2,$$
$$v = -v - 2x^2 y.$$
(4.5)

Now, we can observe that the first two equations in (4.5) can be obtained from the last four. Thus, we can drop the first two equations and still have the equivalent system. Doing that gives:

$$\begin{aligned}\dot{x} &= v,\\ \dot{y} &= w,\\ \dot{w} &= -w + 2xy^2,\\ \dot{v} &= -v - 2x^2 y.\end{aligned} \qquad (4.6)$$

Simplifying the last two equations gives:

$$\begin{aligned}\dot{x} &= v,\\ \dot{y} &= w,\\ w &= xy^2,\\ v &= -x^2 y.\end{aligned} \qquad (4.7)$$

It is important to realize that the equations (4.7) are completely equivalent to the original Euler-Lagrange equations (4.3).

The last two equations in (4.7) do not contain time derivatives. So, they give us conditions that relate positions and velocities of any solutions of the Euler-Lagrange equations at any moment in time. We will call such conditions constraints.

Since constraints must hold at any time, time derivatives of both sides of them must be equal. So, let us take the time derivative of the constraints in the system (4.7). They give us new equations, which we can write together with the rest of the system, getting:

$$\begin{aligned}\dot{x} &= v,\\ \dot{y} &= w,\\ w &= xy^2,\\ v &= -x^2 y,\\ \dot{w} &= \dot{x}y^2 + 2xy\dot{y},\\ \dot{v} &= -2x\dot{x}y - x^2\dot{y}.\end{aligned} \qquad (4.8)$$

Now we use the first two equations in (4.8) to get rid of time derivatives on the right sides of the last two equations, getting:

$$\dot{x} = v,$$
$$\dot{y} = w,$$
$$w = xy^2,$$
$$v = -x^2 y, \tag{4.9}$$
$$\dot{w} = vy^2 + 2xyw,$$
$$\dot{v} = -2xvy - x^2 w.$$

Finally, using the constraints, we eliminate the velocities from the right sides of the last two equations in (4.9), and change the order of the equations. We get:

$$\dot{v} = x^3 y^2,$$
$$\dot{w} = x^2 y^3,$$
$$\dot{x} = v,$$
$$\dot{y} = w, \tag{4.10}$$
$$v = -x^2 y,$$
$$w = xy^2.$$

It is important to realize that the equations (4.10) are completely equivalent to the original Euler-Lagrange equations (4.3). We still call (4.10) the Euler-Lagrange equations.

Equations (4.10) contain the expressions for the time derivatives of all variables. They also give all the constraints of the system. In this sense, they are complete. We just finished what is called the completion process of the original Euler-Lagrange equations (4.3).

Notice that the constraints appeared even before we tried to construct the Hamiltonian. They are already present in the Lagrange formalism; they are consequences of the original Euler-Lagrange equations (4.2).

2. An Example of a Constrained Lagrangian, the Legendre Transformation

Following the definition (3.1) in Chapter III, we introduce the position-momenta variables as

$$x = x,$$
$$y = y,$$
$$p_x = \frac{\partial L}{\partial v}, \qquad (4.11)$$
$$p_y = \frac{\partial L}{\partial w}.$$

Calculating the specific form of the partial derivatives of the Lagrangian (4.1) and placing them into (4.11), we get:

$$x = x,$$
$$y = y,$$
$$p_x = -y, \qquad (4.12)$$
$$p_y = x.$$

The (4.12) is the Legendre Transformation, with the old (x, y, v, w) variables on the right side of (4.12) and the new variables (x, y, p_x, p_y) on the left side. Since the equations (4.12) do not contain the velocities v or w, they give no chance to express them by (x, y, p_x, p_y). The transformation (4.12) is not invertible. So, the Legendre Transformation for the Lagrangian (4.1) is not invertible.

In Chapter III, we defined a Lagrangian to be hyperregular if its Legendre Transformation is invertible. So, the Lagrangian (4.1) is non-hyperregular. Another word for non-hyperregular is "constrained." Therefore, the Lagrangian (4.1) is constrained.

3. An Example of a Constrained Lagrangian, the Hamilton's Equations of Motion

As we said before, the Hamilton's equations are the Euler-Lagrange equations expressed in the position-momenta variables. In the case of hyperregular Lagrangians, the velocities are given directly as functions of positions and momenta (3.2). So, we can simply replace all velocities in the Euler-Lagrange equations with the formulas from (3.2), and the process is essentially finished. We have the Hamilton's equations.

The process is not as simple in the case of a constrained Lagrangian since, by definition, such a Lagrangian does not give us the formulas expressing velocities by positions and momenta; at least it does not give them to us directly. So, we need to do some extra work to get the result. Here we will show how it can be done for the specific Lagrangian (4.1). We will discuss a general procedure in later chapters.

We start by combining the Euler-Lagrange equations (4.2) for the Lagrangian (4.1) with the Legendre Transformation (4.11). We get:

$$\begin{aligned}
\frac{d}{dt}\left[\frac{\partial L}{\partial v}\right] &= \frac{\partial L}{\partial x}, \\
\frac{d}{dt}\left[\frac{\partial L}{\partial w}\right] &= \frac{\partial L}{\partial y}, \\
\frac{dx}{dt} &= v, \\
\frac{dy}{dt} &= w, \\
x &= x, \\
y &= y, \\
p_x &= \frac{\partial L}{\partial v}, \\
p_y &= \frac{\partial L}{\partial w}.
\end{aligned} \qquad (4.13)$$

We aim to manipulate these equations to get the time derivatives of positions and momenta expressed by positions and momenta only. We start by dropping fifth and sixth equations in (4.13) as trivial. Then we replace the partial derivatives on the left side of the first two equations with the momenta from the last two equations. We also change the $\frac{d}{dt}$ symbol with a dot above the variables. We get:

$$\dot{p}_x = \frac{\partial L}{\partial x},$$
$$\dot{p}_y = \frac{\partial L}{\partial y},$$
$$\dot{x} = v,$$
$$\dot{y} = w,$$
$$p_x = \frac{\partial L}{\partial v},$$
$$p_y = \frac{\partial L}{\partial w}.$$

(4.14)

Then we calculate the partial derivatives of the Lagrangian (4.1), and place them in (4.14), getting:

$$\dot{p}_x = w - 2xy^2,$$
$$\dot{p}_y = -v - 2x^2y,$$
$$\dot{x} = v,$$
$$\dot{y} = w,$$
$$p_x = -y,$$
$$p_y = x.$$

(4.15)

The last two equations contain no time derivatives; therefore, we call them constraints. They must hold for all times, so the time derivatives of the left side of each must be equal to the time derivative of the right side. Taking these time derivatives, we get:

$$\dot{p}_x = w - 2xy^2,$$
$$\dot{p}_y = -v - 2x^2y,$$
$$\dot{x} = v,$$
$$\dot{y} = w,$$
$$p_x = -y,$$
$$p_y = x,$$
$$\dot{p}_x = -\dot{y},$$
$$\dot{p}_y = \dot{x}.$$

(4.16)

Then we replace the time derivatives in the last two equations by the right sides of the first four equations, getting:

$$\begin{aligned}
\dot{p}_x &= w - 2xy^2, \\
\dot{p}_y &= -v - 2x^2 y, \\
\dot{x} &= v, \\
\dot{y} &= w, \\
p_x &= -y, \\
p_y &= x, \\
\dot{p}_x &= -\dot{y}, \\
\dot{p}_y &= \dot{x}, \\
w - 2xy^2 &= -w, \\
-v - 2x^2 y &= v.
\end{aligned} \quad (4.17)$$

We simplify the last two equations, getting:

$$\begin{aligned}
\dot{p}_x &= w - 2xy^2, \\
\dot{p}_y &= -v - 2x^2 y, \\
\dot{x} &= v, \\
\dot{y} &= w, \\
p_x &= -y, \\
p_y &= x, \\
\dot{p}_x &= -\dot{y}, \\
\dot{p}_y &= \dot{x}, \\
w &= xy^2, \\
v &= -x^2 y.
\end{aligned} \quad (4.18)$$

Now we replace v and w in the first four equations by the expressions given in the last two. We get:

$$\begin{aligned}
\dot{p}_x &= xy^2 - 2xy^2, \\
\dot{p}_y &= x^2 y - 2x^2 y,
\end{aligned}$$

(continued on the next page)

$$\dot{x} = -x^2 y,$$
$$\dot{y} = xy^2,$$
$$p_x = -y,$$
$$p_y = x,$$
$$\dot{p}_x = -\dot{y},$$
$$\dot{p}_y = \dot{x},$$
$$v = -x^2 y,$$
$$w = xy^2.$$
(4.19)

We simplify the right sides of the first two equations, getting:

$$\dot{p}_x = -xy^2,$$
$$\dot{p}_y = -x^2 y,$$
$$\dot{x} = -x^2 y,$$
$$\dot{y} = xy^2,$$
$$p_x = -y,$$
$$p_y = x,$$
$$\dot{p}_x = -\dot{y},$$
$$\dot{p}_y = \dot{x},$$
$$v = -x^2 y,$$
$$w = xy^2.$$
(4.20)

Then we replace time derivatives on the right sides of the seventh and eighth equations by the right sides of the third and fourth equations, getting:

$$\dot{p}_x = -xy^2,$$
$$\dot{p}_y = -x^2 y,$$
$$\dot{x} = -x^2 y,$$
$$\dot{y} = xy^2,$$

(continued on the next page)

$$p_x = -y,$$
$$p_y = x,$$
$$\dot{p}_x = -xy^2,$$
$$\dot{p}_y = -x^2 y, \qquad (4.21)$$
$$v = -x^2 y,$$
$$w = xy^2.$$

Then we realize that equations one and two are identical to equations seven and eight respectively, so the latter can be dropped from the system, giving:

$$\dot{p}_x = -xy^2,$$
$$\dot{p}_y = -x^2 y,$$
$$\dot{x} = -x^2 y,$$
$$\dot{y} = xy^2,$$
$$p_x = -y, \qquad (4.22)$$
$$p_y = x,$$
$$v = -x^2 y,$$
$$w = xy^2.$$

The above process, in which we obtained new time derivatives of some variables, was another example of the completion process.

Finally, since the Hamilton's equations are supposed to only deal with the equations that provide time derivatives of the position-momenta variables and the relations between position-momenta variables, we drop the last two equations, getting:

$$\dot{p}_x = -xy^2,$$
$$\dot{p}_y = -x^2 y,$$
$$\dot{x} = -x^2 y,$$
$$\dot{y} = xy^2,$$

(continued on the next page)

$$p_x = -y,$$
$$p_y = x. \qquad (4.23)$$

Notice that the complete information about the dropped equations is still contained in (4.23) because the dropped velocities are equal, by definition, to the time derivatives of positions given by equations three and four in (4.23).

The equations (4.23) are the Hamilton's equations for the Lagrangian (4.1). The last two of them, since they do not contain the time derivatives, are constraints. These equations are equivalent to the Euler-Lagrange equations (4.10).

4. An Example of a Constrained Lagrangian, the Hamiltonian

The Hamiltonian in the hyperregular case was introduced by (3.7) in Chapter III, as:

$$H(q,p) = \sum_{i=1}^{N} p_i \cdot v_i(q,p) - L(q,v(q,p)).$$

For the Lagrangian (4.1), we get:

$$H(x,y,p_x,p_y) = p_x v + p_y w - \left(xw - yv - x^2 y^2\right), \qquad (4.24)$$

which can be simplified to:

$$H(x,y,p_x,p_y) = p_x v + p_y w - xw + yv + x^2 y^2. \qquad (4.25)$$

The main reason to calculate a Hamiltonian is to get the Hamilton's equations (3.14) from Chapter III. These equations require using the partial derivatives in position-momenta variables. This, in turn, requires elimination of velocities from the Hamiltonian (or at least the knowledge of velocities expressed by positions and momenta so we can use the chain rule).

In the case of the hyperregular Lagrangian, we had the inverse of the Legendre Transformations (3.2). This inverse would allow us to express velocities by positions and momenta, so that velocities could be removed from the Hamiltonian. In the case of the Lagrangian (4.1), the Legendre Transformations (4.12) would not give us expressions for velocities in terms of positions and momenta.

So, following what worked for the hyperregular Lagrangian is not possible here. Therefore, we need to decide how to proceed.

One possibility is to notice that if we factor velocities in the Hamiltonian (4.24) as common factors we get:

$$H(x, y, p_x, p_y) = v(p_x + y) + w(p_y - x) + x^2 y^2. \tag{4.26}$$

We still have the velocities in (4.26), but now we can use the constraints $p_x = -y$, and $p_y = x$ from the equations (4.15), to replace momenta by positions in (4.26). We get:

$$H_1(x, y, p_x, p_y) = v(p_x + y) + w(p_y - x) + x^2 y^2 =$$
$$= v(-y + y) + w(x - x) + x^2 y^2 =$$
$$= v \cdot 0 + w \cdot 0 + x^2 y^2 = x^2 y^2$$

So, we obtained the Hamiltonian without explicit velocities, as we wanted:

$$H_1(x, y, p_x, p_y) = x^2 y^2 \tag{4.27}$$

We called this Hamiltonian H_1, because we will also be calculating different Hamiltonians. In a moment we will see that the Hamiltonian H_1 is not a good Hamiltonian for the Lagrangian (4.1).

The partial derivatives of the Hamiltonian in (4.27) are:

$$\frac{\partial H_1}{\partial x} = 2xy^2,$$
$$\frac{\partial H_1}{\partial y} = 2x^2 y,$$

(continued on the next page)

$$\frac{\partial H_1}{\partial p_x} = 0,$$

$$\frac{\partial H_1}{\partial p_y} = 0.$$

Therefore, the Hamilton's equations (3.14) would become:

$$\begin{aligned}
\dot{p}_x &= -\frac{\partial H_1}{\partial x} = -2xy^2, \\
\dot{p}_y &= -\frac{\partial H_1}{\partial y} = -2x^2 y, \\
\dot{x} &= \frac{\partial H_1}{\partial p_x} = 0, \\
\dot{y} &= \frac{\partial H_1}{\partial p_y} = 0.
\end{aligned} \quad (4.28)$$

However, the equations (4.28) are not what is expected here. By comparing the equations (4.28) and (4.23), we can see that they are different. The difference cannot be accommodated by using the constraints, since the time derivatives of positions are zero in equations (4.28) and non-zero in (4.23). The conclusion is, the Hamiltonian H_1 from (4.27) is not reproducing the Euler-Lagrange equations (4.4), nor their equivalence, the equations (4.23). Therefore H_1 is not acceptable as a Hamiltonian for our system.

Let us try something else then. Notice that in the process of getting Hamilton's equations from the Euler-Lagrange equations, we got the expressions for velocities in (4.22). These expressions were:

$$\begin{aligned}
v &= -x^2 y, \\
w &= xy^2.
\end{aligned} \quad (4.29)$$

We can try to place them in the expression (4.25) for the Hamiltonian to get rid of velocities. We get:

$$H_2(x, y, p_x, p_y) = p_x v + p_y w - xw + yv + x^2 y^2 =$$
$$= p_x(-x^2 y) + p_y(xy^2) - x(xy^2) + y(-x^2 y) + x^2 y^2 =$$
$$= -p_x x^2 y + p_y xy^2 - x^2 y^2 - x^2 y^2 + x^2 y^2 =$$
$$= -p_x x^2 y + p_y xy^2 - x^2 y^2.$$

So, this gives the Hamiltonian as:

$$H_2(x, y, p_x, p_y) = -p_x x^2 y + p_y xy^2 - x^2 y^2. \tag{4.30}$$

The partial derivatives of the Hamiltonian H_2 from (4.30) are:

$$\frac{\partial H_2}{\partial x} = -2p_x xy + p_y y^2 - 2xy^2,$$
$$\frac{\partial H_2}{\partial y} = -p_x x^2 + 2p_y xy - 2x^2 y,$$
$$\frac{\partial H_2}{\partial p_x} = -x^2 y, \tag{4.31}$$
$$\frac{\partial H_2}{\partial p_y} = xy^2.$$

We can use the constraints $p_x = -y$, and $p_y = x$ from the Legendre Transformations (4.12), to replace momenta by positions in (4.31), getting:

$$\frac{\partial H_2}{\partial x} = 2yxy + xy^2 - 2xy^2,$$
$$\frac{\partial H_2}{\partial y} = yx^2 + 2xxy - 2x^2 y,$$
$$\frac{\partial H_2}{\partial p_x} = -x^2 y, \tag{4.32}$$
$$\frac{\partial H_2}{\partial p_y} = xy^2.$$

Simplifying this we get:

$$\frac{\partial H_2}{\partial x} = xy^2,$$

$$\frac{\partial H_2}{\partial y} = x^2 y,$$

$$\frac{\partial H_2}{\partial p_x} = -x^2 y, \tag{4.33}$$

$$\frac{\partial H_2}{\partial p_y} = xy^2.$$

Placing the partial derivatives (4.33) in the Hamilton's equations (3.14) gives:

$$\dot{p}_x = -\frac{\partial H_2}{\partial x} = -xy^2,$$

$$\dot{p}_y = -\frac{\partial H_2}{\partial y} = -x^2 y,$$

$$\dot{x} = \frac{\partial H_2}{\partial p_x} = -x^2 y, \tag{4.34}$$

$$\dot{y} = \frac{\partial H_2}{\partial p_y} = xy^2.$$

The equations (4.34) are identical to the first four equations of (4.23). This means that the Hamiltonian obtained by eliminating the velocities using the velocities obtained during the process of deriving the Hamilton's equations, namely (4.30), is the correct Hamiltonian. A rather obvious hypothesis would be that we should always use velocities from that process. However, the hypothesis does not look promising. Notice that when getting the velocities, we can always use constraints to change the formulas for them, and this would give us a different Hamiltonian. So, it seems that this approach works here by chance only.

Let us try another form of a possible Hamiltonian. We can start with the one that we already established as incorrect, namely $H_1(x, y, p_x, p_y) = x^2 y^2$ from (4.27), and then replace one x and one y in it using the $p_x = -y$, and $p_y = x$ from the equations (4.15). We get:

$$H_3\left(x, y, p_x, p_y\right) = -p_x p_y xy. \tag{4.35}$$

The partial derivatives of the Hamiltonian (4.35) are:

$$\frac{\partial H_3}{\partial x} = -p_x p_y y,$$

$$\frac{\partial H_3}{\partial y} = -p_x p_y x,$$

$$\frac{\partial H_3}{\partial p_x} = -p_y xy,$$

$$\frac{\partial H_3}{\partial p_y} = -p_x xy.$$

Using the constraints $p_x = -y$, and $p_y = x$ from (4.15), we can rewrite these derivatives as:

$$\begin{aligned}\frac{\partial H_3}{\partial x} &= xy^2, \\ \frac{\partial H_3}{\partial y} &= x^2 y, \\ \frac{\partial H_3}{\partial p_x} &= -x^2 y, \\ \frac{\partial H_3}{\partial p_y} &= xy^2.\end{aligned} \tag{4.36}$$

Placing the partial derivatives (4.36) in the Hamilton's equations (3.14) gives:

$$\dot{p}_x = -\frac{\partial H_3}{\partial x} = -xy^2,$$

$$\dot{p}_y = -\frac{\partial H_3}{\partial y} = -x^2 y,$$

(continued on the next page)

$$\dot{x} = \frac{\partial H_3}{\partial p_x} = -x^2 y,$$
$$\dot{y} = \frac{\partial H_3}{\partial p_y} = xy^2. \qquad (4.37)$$

The equations (4.37) are identical to the first four equations of (4.23), This means that the Hamiltonian (4.35), which was essentially obtained by guessing, is also a correct Hamiltonian.

So, we may conclude that many correct Hamiltonians exist, which differ by using different combinations of substitutions of constraints. We should also observe that not every substitution of the constraints leads to getting a correct Hamiltonian.

In addition, we should conclude that we need a more general approach to substituting different combinations of constraints into expressions. Later we will devote an entire chapter to describing substitutions in a general way.

5. An Example of a Constrained Lagrangian, the Poisson Brackets and the Hamilton's Equations of Motion

For constrained systems, we use the same definition (3.19) from Chapter III of the Poisson Brackets that we used for the hyperregular systems, namely:

$$\{F,G\} = \sum_{i=1}^{N} \left(\frac{\partial F}{\partial q_i} \cdot \frac{\partial G}{\partial p_i} - \frac{\partial F}{\partial p_i} \cdot \frac{\partial G}{\partial q_i} \right). \qquad (4.38)$$

For this example, we get:

$$\{F,G\} = \frac{\partial F}{\partial x} \cdot \frac{\partial G}{\partial p_x} - \frac{\partial F}{\partial p_x} \cdot \frac{\partial G}{\partial x} + \frac{\partial F}{\partial y} \cdot \frac{\partial G}{\partial p_y} - \frac{\partial F}{\partial p_y} \cdot \frac{\partial G}{\partial y}. \qquad (4.39)$$

Since earlier we got the usual and correct Hamilton's equations for both the Hamiltonian H_2 (4.30) and the Hamiltonian H_3 (4.35), we also got the usual and correct Hamilton's equations expressed by the Poisson Brackets (4.39), namely:

$$\dot{p}_x = \{p_x, H_2\},$$
$$\dot{p}_y = \{p_y, H_2\},$$
$$\dot{x} = \{x, H_2\},$$
$$\dot{y} = \{y, H_2\}, \qquad (4.40)$$
$$p_x = -y,$$
$$p_y = x,$$

or

$$\dot{p}_x = \{p_x, H_3\},$$
$$\dot{p}_y = \{p_y, H_3\},$$
$$\dot{x} = \{x, H_3\},$$
$$\dot{y} = \{y, H_3\}, \qquad (4.41)$$
$$p_x = -y,$$
$$p_y = x.$$

The last two equations in (4.40) and (4.41) are the constraints from (4.23). The constraints must be included in the Hamilton's equations, or the Hamilton's equations would not be equivalent to the starting Euler-Lagrange equations.

6. An Example of a Constrained Lagrangian, the Newtonian Equations of Motion

The simplest way to obtain the Newtonian equations of motion for this example is to go back to the equations (4.10), namely:

$$\dot{v} = x^3 y^2,$$
$$\dot{w} = x^2 y^3,$$
$$\dot{x} = v,$$
$$\dot{y} = w, \qquad (4.42)$$
$$v = -x^2 y,$$
$$w = xy^2.$$

The equations (4.42) are the Newtonian equations for the constrained Lagrangian (4.1).

Replacing the velocities in the equations in the third and fourth equations in (4.42) using the equation five and six, the Newtonian equations (4.42) can be rewritten as a system of the second order, namely:

$$\begin{aligned}\ddot{x} &= x^3 y^2, \\ \ddot{y} &= x^2 y^3, \\ \dot{x} &= -x^2 y, \\ \dot{y} &= xy^2.\end{aligned} \quad (4.43)$$

The equations three and four in the equations (4.42) may then be merely looked at as somewhat irrelevant definitions of velocities that we do not have to use. Traditionally, when considering the Newtonian Equations, we use the symbols $\dot{x}$ and $\dot{y}$ rather than v and w.

7. An Example of a Constrained Lagrangian, the Dirac's Brackets

The Dirac's Brackets will be fully introduced in Chapter XIII. Here, we will just calculate them without too many explanations.

We start with defining the constraints φ_1 and φ_2 of the system, using (x, y, p_x, p_y) variables, as expressions equal to zero. Using the last two equations in (4.40), and combining all terms on one side of the equations, we get:

$$\begin{aligned}\varphi_1 &= x - p_y = 0, \\ \varphi_2 &= y + p_x = 0.\end{aligned} \quad (4.44)$$

Then we calculate the Poisson Bracket, using the formula (4.39):

$$\{\varphi_1, \varphi_2\} = \frac{\partial \varphi_1}{\partial x} \cdot \frac{\partial \varphi_2}{\partial p_x} - \frac{\partial \varphi_1}{\partial p_x} \cdot \frac{\partial \varphi_2}{\partial x} + \frac{\partial \varphi_1}{\partial y} \cdot \frac{\partial \varphi_2}{\partial p_y} - \frac{\partial \varphi_1}{\partial p_y} \cdot \frac{\partial \varphi_2}{\partial y} = \\ = 1 \cdot 1 - 0 \cdot 0 + 0 \cdot 0 - (-1) \cdot 1 = 2. \quad (4.45)$$

So, we have:

$$\{\varphi_1, \varphi_2\} = 2. \tag{4.46}$$

Following Dirac (check Chapter XIII for more explanations), we define the constraint matrix C^{-1} as:

$$C^{-1} = \begin{pmatrix} 0 & \{\varphi_1, \varphi_2\} \\ -\{\varphi_1, \varphi_2\} & 0 \end{pmatrix} = \begin{pmatrix} 0 & 2 \\ -2 & 0 \end{pmatrix}. \tag{4.47}$$

The inverse of this matrix is:

$$C = \begin{pmatrix} 0 & -\dfrac{1}{2} \\ \dfrac{1}{2} & 0 \end{pmatrix}. \tag{4.48}$$

The Dirac's Brackets are then defined using the Poisson Brackets and the matrix C as:

$$\{F, G\}_D = \{F, G\} - (\{F, \varphi_1\}, \{F, \varphi_2\}) \begin{pmatrix} 0 & -\dfrac{1}{2} \\ \dfrac{1}{2} & 0 \end{pmatrix} \begin{pmatrix} \{\varphi_1, G\} \\ \{\varphi_2, G\} \end{pmatrix}. \tag{4.49}$$

Using (4.39), we can calculate pieces needed for the calculations inside of (4.49) as:

$$\{F, \varphi_1\} = \frac{\partial F}{\partial x} \cdot \frac{\partial \varphi_1}{\partial p_x} - \frac{\partial F}{\partial p_x} \cdot \frac{\partial \varphi_1}{\partial x} + \frac{\partial F}{\partial y} \cdot \frac{\partial \varphi_1}{\partial p_y} - \frac{\partial F}{\partial p_y} \cdot \frac{\partial \varphi_1}{\partial y} =$$

$$= \frac{\partial F}{\partial x} \cdot 0 - \frac{\partial F}{\partial p_x} \cdot 1 + \frac{\partial F}{\partial y} \cdot (-1) - \frac{\partial F}{\partial p_y} \cdot 0 = -\frac{\partial F}{\partial p_x} - \frac{\partial F}{\partial y},$$

so, we get:

$$\{F, \varphi_1\} = -\frac{\partial F}{\partial p_x} - \frac{\partial F}{\partial y}.$$

Then:

$$\{F, \varphi_2\} = \frac{\partial F}{\partial x} \cdot \frac{\partial \varphi_2}{\partial p_x} - \frac{\partial F}{\partial p_x} \cdot \frac{\partial \varphi_2}{\partial x} + \frac{\partial F}{\partial y} \cdot \frac{\partial \varphi_2}{\partial p_y} - \frac{\partial F}{\partial p_y} \cdot \frac{\partial \varphi_2}{\partial y} =$$

$$= \frac{\partial F}{\partial x} \cdot 1 - \frac{\partial F}{\partial p_x} \cdot 0 + \frac{\partial F}{\partial y} \cdot 0 - \frac{\partial F}{\partial p_y} \cdot 1 = \frac{\partial F}{\partial x} - \frac{\partial F}{\partial p_y},$$

so, we get:

$$\{F, \varphi_2\} = \frac{\partial F}{\partial x} - \frac{\partial F}{\partial p_y},$$

Then:

$$\{\varphi_1, G\} = \frac{\partial \varphi_1}{\partial x} \cdot \frac{\partial G}{\partial p_x} - \frac{\partial \varphi_1}{\partial p_x} \cdot \frac{\partial G}{\partial x} + \frac{\partial \varphi_1}{\partial y} \cdot \frac{\partial G}{\partial p_y} - \frac{\partial \varphi_1}{\partial p_y} \cdot \frac{\partial G}{\partial y} =$$

$$= 1 \cdot \frac{\partial G}{\partial p_x} - 0 \cdot \frac{\partial G}{\partial x} + 0 \cdot \frac{\partial G}{\partial p_y} - (-1) \cdot \frac{\partial G}{\partial y} = \frac{\partial G}{\partial p_x} + \frac{\partial G}{\partial y},$$

so, we get:

$$\{\varphi_1, G\} = \frac{\partial G}{\partial p_x} + \frac{\partial G}{\partial y}.$$

Then:

$$\{\varphi_2, G\} = \frac{\partial \varphi_2}{\partial x} \cdot \frac{\partial G}{\partial p_x} - \frac{\partial \varphi_2}{\partial p_x} \cdot \frac{\partial G}{\partial x} + \frac{\partial \varphi_2}{\partial y} \cdot \frac{\partial G}{\partial p_y} - \frac{\partial \varphi_2}{\partial p_y} \cdot \frac{\partial G}{\partial y} =$$

$$= 0 \cdot \frac{\partial G}{\partial p_x} - 1 \cdot \frac{\partial G}{\partial x} + 1 \cdot \frac{\partial G}{\partial p_y} - 0 \cdot \frac{\partial G}{\partial y} = -\frac{\partial G}{\partial x} + \frac{\partial G}{\partial p_y},$$

so, we get:

$$\{\varphi_2, G\} = -\frac{\partial G}{\partial x} + \frac{\partial G}{\partial p_y}.$$

Placing the results obtained above into the (4.49), and using (4.39), we get:

$$\{F,G\}_D = \{F,G\} - (\{F,\varphi_1\},\{F,\varphi_2\}) \begin{pmatrix} 0 & -\frac{1}{2} \\ \frac{1}{2} & 0 \end{pmatrix} \begin{pmatrix} \{\varphi_1,G\} \\ \{\varphi_2,G\} \end{pmatrix} =$$

$$= \{F,G\} = \frac{\partial F}{\partial x} \cdot \frac{\partial G}{\partial p_x} - \frac{\partial F}{\partial p_x} \cdot \frac{\partial G}{\partial x} + \frac{\partial F}{\partial y} \cdot \frac{\partial G}{\partial p_y} - \frac{\partial F}{\partial p_y} \cdot \frac{\partial G}{\partial y} +$$

$$- \left(-\frac{\partial F}{\partial p_x} - \frac{\partial F}{\partial y}, \frac{\partial F}{\partial x} - \frac{\partial F}{\partial p_y} \right) \begin{pmatrix} 0 & -\frac{1}{2} \\ \frac{1}{2} & 0 \end{pmatrix} \begin{pmatrix} \frac{\partial G}{\partial p_x} + \frac{\partial G}{\partial y} \\ -\frac{\partial G}{\partial x} + \frac{\partial G}{\partial p_y} \end{pmatrix} =$$

$$= \frac{\partial F}{\partial x} \cdot \frac{\partial G}{\partial p_x} - \frac{\partial F}{\partial p_x} \cdot \frac{\partial G}{\partial x} + \frac{\partial F}{\partial y} \cdot \frac{\partial G}{\partial p_y} - \frac{\partial F}{\partial p_y} \cdot \frac{\partial G}{\partial y} +$$

$$-\frac{1}{2}\left(-\frac{\partial F}{\partial p_x} - \frac{\partial F}{\partial y}, \frac{\partial F}{\partial x} - \frac{\partial F}{\partial p_y} \right) \begin{pmatrix} \frac{\partial G}{\partial x} - \frac{\partial G}{\partial p_y} \\ \frac{\partial G}{\partial p_x} + \frac{\partial G}{\partial y} \end{pmatrix} =$$

$$= \frac{\partial F}{\partial x} \cdot \frac{\partial G}{\partial p_x} - \frac{\partial F}{\partial p_x} \cdot \frac{\partial G}{\partial x} + \frac{\partial F}{\partial y} \cdot \frac{\partial G}{\partial p_y} - \frac{\partial F}{\partial p_y} \cdot \frac{\partial G}{\partial y} +$$

$$-\frac{1}{2}\left(-\frac{\partial F}{\partial p_x} - \frac{\partial F}{\partial y} \right)\left(\frac{\partial G}{\partial x} - \frac{\partial G}{\partial p_y} \right) - \frac{1}{2}\left(\frac{\partial F}{\partial x} - \frac{\partial F}{\partial p_y} \right)\left(\frac{\partial G}{\partial p_x} + \frac{\partial G}{\partial y} \right) =$$

$$= \frac{\partial F}{\partial x} \cdot \frac{\partial G}{\partial p_x} - \frac{\partial F}{\partial p_x} \cdot \frac{\partial G}{\partial x} + \frac{\partial F}{\partial y} \cdot \frac{\partial G}{\partial p_y} - \frac{\partial F}{\partial p_y} \cdot \frac{\partial G}{\partial y} +$$

$$-\frac{1}{2}\left(-\frac{\partial F}{\partial p_x} \right)\left(\frac{\partial G}{\partial x} \right) - \frac{1}{2}\left(-\frac{\partial F}{\partial p_x} \right)\left(-\frac{\partial G}{\partial p_y} \right) - \frac{1}{2}\left(-\frac{\partial F}{\partial y} \right)\left(\frac{\partial G}{\partial x} \right) - \frac{1}{2}\left(-\frac{\partial F}{\partial y} \right)\left(-\frac{\partial G}{\partial p_y} \right) +$$

$$-\frac{1}{2}\left(\frac{\partial F}{\partial x} \right)\left(\frac{\partial G}{\partial p_x} \right) - \frac{1}{2}\left(\frac{\partial F}{\partial x} \right)\left(\frac{\partial G}{\partial y} \right) - \frac{1}{2}\left(-\frac{\partial F}{\partial p_y} \right)\left(\frac{\partial G}{\partial p_x} \right) - \frac{1}{2}\left(-\frac{\partial F}{\partial p_y} \right)\left(\frac{\partial G}{\partial y} \right) =$$

$$= \frac{\partial F}{\partial x} \cdot \frac{\partial G}{\partial p_x} - \frac{\partial F}{\partial p_x} \cdot \frac{\partial G}{\partial x} + \frac{\partial F}{\partial y} \cdot \frac{\partial G}{\partial p_y} - \frac{\partial F}{\partial p_y} \cdot \frac{\partial G}{\partial y} +$$

$$+\frac{1}{2} \cdot \frac{\partial F}{\partial p_x} \cdot \frac{\partial G}{\partial x} - \frac{1}{2} \cdot \frac{\partial F}{\partial p_x} \cdot \frac{\partial G}{\partial p_y} + \frac{1}{2} \cdot \frac{\partial F}{\partial y} \cdot \frac{\partial G}{\partial x} - \frac{1}{2} \cdot \frac{\partial F}{\partial y} \cdot \frac{\partial G}{\partial p_y} +$$

(continued on the next page)

$$-\frac{1}{2}\cdot\frac{\partial F}{\partial x}\cdot\frac{\partial G}{\partial p_x} - \frac{1}{2}\cdot\frac{\partial F}{\partial x}\cdot\frac{\partial G}{\partial y} + \frac{1}{2}\cdot\frac{\partial F}{\partial p_y}\cdot\frac{\partial G}{\partial p_x} + \frac{1}{2}\cdot\frac{\partial F}{\partial p_y}\cdot\frac{\partial G}{\partial y} =$$

$$= \frac{1}{2}\cdot\frac{\partial F}{\partial x}\cdot\frac{\partial G}{\partial p_x} - \frac{1}{2}\cdot\frac{\partial F}{\partial p_x}\cdot\frac{\partial G}{\partial x} + \frac{1}{2}\cdot\frac{\partial F}{\partial y}\cdot\frac{\partial G}{\partial p_y} - \frac{1}{2}\cdot\frac{\partial F}{\partial p_y}\cdot\frac{\partial G}{\partial y} +$$

$$-\frac{1}{2}\cdot\frac{\partial F}{\partial p_x}\cdot\frac{\partial G}{\partial p_y} + \frac{1}{2}\cdot\frac{\partial F}{\partial y}\cdot\frac{\partial G}{\partial x} - \frac{1}{2}\cdot\frac{\partial F}{\partial x}\cdot\frac{\partial G}{\partial y} + \frac{1}{2}\cdot\frac{\partial F}{\partial p_y}\cdot\frac{\partial G}{\partial p_x}.$$

So, the Dirac's Brackets are given as:

$$\{F,G\}_D = \frac{1}{2}\cdot\frac{\partial F}{\partial x}\cdot\frac{\partial G}{\partial p_x} - \frac{1}{2}\cdot\frac{\partial F}{\partial p_x}\cdot\frac{\partial G}{\partial x} + \frac{1}{2}\cdot\frac{\partial F}{\partial y}\cdot\frac{\partial G}{\partial p_y} - \frac{1}{2}\cdot\frac{\partial F}{\partial p_y}\cdot\frac{\partial G}{\partial y} +$$
$$-\frac{1}{2}\cdot\frac{\partial F}{\partial p_x}\cdot\frac{\partial G}{\partial p_y} + \frac{1}{2}\cdot\frac{\partial F}{\partial y}\cdot\frac{\partial G}{\partial x} - \frac{1}{2}\cdot\frac{\partial F}{\partial x}\cdot\frac{\partial G}{\partial y} + \frac{1}{2}\cdot\frac{\partial F}{\partial p_y}\cdot\frac{\partial G}{\partial p_x}. \tag{4.50}$$

8. An Example of a Constrained Lagrangian, the Hamiltonian H_1 and the Dirac's Brackets

Let us calculate the Dirac's Brackets of the Hamiltonian H_1 and the variables (x, y, p_x, p_y). The Dirac's Brackets are given in (4.50). The Hamiltonian H_1 is given in (4.27), as:

$$H_1(x, y, p_x, p_y) = x^2 y^2 \tag{4.51}$$

So, we have:

$$\{x, H\}_D = \frac{1}{2}\cdot\frac{\partial x}{\partial x}\cdot\frac{\partial H_1}{\partial p_x} - \frac{1}{2}\cdot\frac{\partial x}{\partial p_x}\cdot\frac{\partial H_1}{\partial x} + \frac{1}{2}\cdot\frac{\partial x}{\partial y}\cdot\frac{\partial H_1}{\partial p_y} - \frac{1}{2}\cdot\frac{\partial x}{\partial p_y}\cdot\frac{\partial H_1}{\partial y} +$$

$$-\frac{1}{2}\cdot\frac{\partial x}{\partial p_x}\cdot\frac{\partial H_1}{\partial p_y} + \frac{1}{2}\cdot\frac{\partial x}{\partial y}\cdot\frac{\partial H_1}{\partial x} - \frac{1}{2}\cdot\frac{\partial x}{\partial x}\cdot\frac{\partial H_1}{\partial y} + \frac{1}{2}\cdot\frac{\partial x}{\partial p_y}\cdot\frac{\partial H_1}{\partial p_x} =$$

$$= \frac{1}{2}\cdot 1\cdot\frac{\partial H_1}{\partial p_x} - \frac{1}{2}\cdot 0\cdot\frac{\partial H_1}{\partial x} + \frac{1}{2}\cdot 0\cdot\frac{\partial H_1}{\partial p_y} - \frac{1}{2}\cdot 0\cdot\frac{\partial H_1}{\partial y} +$$

$$-\frac{1}{2}\cdot 0\cdot\frac{\partial H_1}{\partial p_y} + \frac{1}{2}\cdot 0\cdot\frac{\partial H_1}{\partial x} - \frac{1}{2}\cdot 1\cdot\frac{\partial H_1}{\partial y} + \frac{1}{2}\cdot 0\cdot\frac{\partial H_1}{\partial p_x} =$$

$$= \frac{1}{2}\cdot\frac{\partial H_1}{\partial p_x} - \frac{1}{2}\cdot\frac{\partial H_1}{\partial y} = \frac{1}{2}\cdot 0 - \frac{1}{2}\cdot 2x^2 y = -x^2 y.$$

So, we get:

$$\{x, H_1\}_D = -x^2 y. \tag{4.52}$$

Then:

$$\{y, H_1\}_D = \frac{1}{2} \cdot \frac{\partial y}{\partial x} \cdot \frac{\partial H_1}{\partial p_x} - \frac{1}{2} \cdot \frac{\partial y}{\partial p_x} \cdot \frac{\partial H_1}{\partial x} + \frac{1}{2} \cdot \frac{\partial y}{\partial y} \cdot \frac{\partial H_1}{\partial p_y} - \frac{1}{2} \cdot \frac{\partial y}{\partial p_y} \cdot \frac{\partial H_1}{\partial y} +$$

$$-\frac{1}{2} \cdot \frac{\partial y}{\partial p_x} \cdot \frac{\partial H_1}{\partial p_y} + \frac{1}{2} \cdot \frac{\partial y}{\partial y} \cdot \frac{\partial H_1}{\partial x} - \frac{1}{2} \cdot \frac{\partial y}{\partial x} \cdot \frac{\partial H_1}{\partial y} + \frac{1}{2} \cdot \frac{\partial y}{\partial p_y} \cdot \frac{\partial H_1}{\partial p_x} =$$

$$= \frac{1}{2} \cdot 0 \cdot \frac{\partial H_1}{\partial p_x} - \frac{1}{2} \cdot 0 \cdot \frac{\partial H_1}{\partial x} + \frac{1}{2} \cdot 1 \cdot \frac{\partial H_1}{\partial p_y} - \frac{1}{2} \cdot 0 \cdot \frac{\partial H_1}{\partial y} +$$

$$-\frac{1}{2} \cdot 0 \cdot \frac{\partial H_1}{\partial p_y} + \frac{1}{2} \cdot 1 \cdot \frac{\partial H_1}{\partial x} - \frac{1}{2} \cdot 0 \cdot \frac{\partial H_1}{\partial y} + \frac{1}{2} \cdot 0 \cdot \frac{\partial H_1}{\partial p_x} =$$

$$= \frac{1}{2} \cdot \frac{\partial H_1}{\partial p_y} + \frac{1}{2} \cdot \frac{\partial H_1}{\partial x} = \frac{1}{2} \cdot 0 + \frac{1}{2} \cdot 2xy^2 = xy^2.$$

So, we get:

$$\{y, H_1\}_D = xy^2. \tag{4.53}$$

Then:

$$\{p_x, H_1\}_D = \frac{1}{2} \cdot \frac{\partial p_x}{\partial x} \cdot \frac{\partial H_1}{\partial p_x} - \frac{1}{2} \cdot \frac{\partial p_x}{\partial p_x} \cdot \frac{\partial H_1}{\partial x} + \frac{1}{2} \cdot \frac{\partial p_x}{\partial y} \cdot \frac{\partial H_1}{\partial p_y} - \frac{1}{2} \cdot \frac{\partial p_x}{\partial p_y} \cdot \frac{\partial H_1}{\partial y} +$$

$$-\frac{1}{2} \cdot \frac{\partial p_x}{\partial p_x} \cdot \frac{\partial H_1}{\partial p_y} + \frac{1}{2} \cdot \frac{\partial p_x}{\partial y} \cdot \frac{\partial H_1}{\partial x} - \frac{1}{2} \cdot \frac{\partial p_x}{\partial x} \cdot \frac{\partial H_1}{\partial y} + \frac{1}{2} \cdot \frac{\partial p_x}{\partial p_y} \cdot \frac{\partial H_1}{\partial p_x} =$$

$$= \frac{1}{2} \cdot 0 \cdot \frac{\partial H_1}{\partial p_x} - \frac{1}{2} \cdot 1 \cdot \frac{\partial H_1}{\partial x} + \frac{1}{2} \cdot 0 \cdot \frac{\partial H_1}{\partial p_y} - \frac{1}{2} \cdot 0 \cdot \frac{\partial H_1}{\partial y} +$$

$$-\frac{1}{2} \cdot 1 \cdot \frac{\partial H_1}{\partial p_y} + \frac{1}{2} \cdot 0 \cdot \frac{\partial H_1}{\partial x} - \frac{1}{2} \cdot 0 \cdot \frac{\partial H_1}{\partial y} + \frac{1}{2} \cdot 0 \cdot \frac{\partial H_1}{\partial p_x} =$$

$$= -\frac{1}{2} \cdot \frac{\partial H_1}{\partial x} - \frac{1}{2} \cdot \frac{\partial H_1}{\partial p_y} = -\frac{1}{2} \cdot 2xy^2 - \frac{1}{2} \cdot 0 = -xy^2.$$

So, we get:

$$\{p_x, H_1\}_D = -xy^2. \tag{4.54}$$

Then:

$$\{p_y, H_1\}_D = \frac{1}{2} \cdot \frac{\partial p_y}{\partial x} \cdot \frac{\partial H_1}{\partial p_x} - \frac{1}{2} \cdot \frac{\partial p_y}{\partial p_x} \cdot \frac{\partial H_1}{\partial x} + \frac{1}{2} \cdot \frac{\partial p_y}{\partial y} \cdot \frac{\partial H_1}{\partial p_y} - \frac{1}{2} \cdot \frac{\partial p_y}{\partial p_y} \cdot \frac{\partial H_1}{\partial y} +$$

$$-\frac{1}{2} \cdot \frac{\partial p_y}{\partial p_x} \cdot \frac{\partial H_1}{\partial p_y} + \frac{1}{2} \cdot \frac{\partial p_y}{\partial y} \cdot \frac{\partial H_1}{\partial x} - \frac{1}{2} \cdot \frac{\partial p_y}{\partial x} \cdot \frac{\partial H_1}{\partial y} + \frac{1}{2} \cdot \frac{\partial p_y}{\partial p_y} \cdot \frac{\partial H_1}{\partial p_x} =$$

$$= \frac{1}{2} \cdot 0 \cdot \frac{\partial H_1}{\partial p_x} - \frac{1}{2} \cdot 0 \cdot \frac{\partial H_1}{\partial x} + \frac{1}{2} \cdot 0 \cdot \frac{\partial H_1}{\partial p_y} - \frac{1}{2} \cdot 1 \cdot \frac{\partial H_1}{\partial y} +$$

$$-\frac{1}{2} \cdot 0 \cdot \frac{\partial H_1}{\partial p_y} + \frac{1}{2} \cdot 0 \cdot \frac{\partial H_1}{\partial x} - \frac{1}{2} \cdot 0 \cdot \frac{\partial H_1}{\partial y} + \frac{1}{2} \cdot 1 \cdot \frac{\partial H_1}{\partial p_x} =$$

$$= -\frac{1}{2} \cdot \frac{\partial H_1}{\partial y} + \frac{1}{2} \cdot \frac{\partial H_1}{\partial p_x} = -\frac{1}{2} \cdot 2x^2 y + \frac{1}{2} \cdot 0 = -x^2 y.$$

So, we get:

$$\{p_y, H_1\}_D = -x^2 y. \tag{4.55}$$

Comparing the equations (4.52), (4.53), (4.54), and (4.55) with the Hamilton's equations (4.23), we get:

$$\begin{aligned}
\dot{p}_x &= \{p_x, H_1\}_D, \\
\dot{p}_y &= \{p_y, H_1\}_D, \\
\dot{x} &= \{x, H_1\}_D, \\
\dot{y} &= \{y, H_1\}_D, \\
p_x &= -y, \\
p_y &= x.
\end{aligned} \tag{4.56}$$

Notice that the Dirac's Brackets with Hamiltonian H_1 give correct equations of motion, while the Poisson Brackets, as we established earlier, do not.

We will skip the calculations, but the reader may check that if we replace H_1 in (4.56) by H_2 or H_3, we will get the same equations of motion. So, any of the Hamiltonians considered earlier will work correctly with the Dirac's Brackets, including the one that was not working correctly with the Poisson Brackets.

We will see in later chapters that Dirac's Brackets have the property that if the constraints used to define the Dirac's Brackets are used to modify the variables by substitutions, then the Dirac's Brackets give the same results for modified and non-modified variables. The reason that all three Hamiltonians give the same Dirac's Brackets with other variables is that all these Hamiltonians can be modified from one to another by substitution that is using the constraints φ_1 and φ_2.

We can also observe that if we decide to use the Dirac's Brackets to generate the equations of motion, instead of the Poisson Brackets, then any Hamiltonian that we obtained in this chapter, becomes a correct Hamiltonian in the sense that it will generate the correct equations of motion.

Later chapters will present what we have shown here in a general setting.

© Hebda 2024

CHAPTER V

FROM THE EULER-LAGRANGE EQUATIONS TO THE HAMILTON'S EQUATIONS AND THE NEWTONIAN EQUATIONS FOR REGULAR CONSTRAINED LAGRANGIANS

In this chapter, we will provide a general process of obtaining the Hamilton's equations of motion and the Newtonian equations of motion from a regular constrained Lagrangian.

1. Regular Constrained Lagrangian, the Hamilton's Equations

As before, let us assume that we have N - dimensional mechanical system described by the variables $(q_1, q_2, ..., q_N, v_1, v_2, ..., v_N)$. Assume also that we have a Lagrangian (2.1):

$$L = L(q_1, q_2, ..., q_N, v_1, v_2, ..., v_N), \tag{5.1}$$

and the equations of motion are the usual Euler-Lagrange equations of motion (2.3):

$$\frac{d}{dt}\left[\frac{\partial L}{\partial v_i}\right] = \frac{\partial L}{\partial q_i},$$
$$\frac{dq_i}{dt} = v_i, \qquad i = 1, ..., N. \tag{5.2}$$

We use the same Legendre Transformations as they were defined in Chapter III by formula (3.1). The Legendre Transformations for constrained systems are identical to what they were for non-constrained systems. They are defined as:

$$q_i = q_i,$$
$$p_i = \frac{\partial L}{\partial v_i}, \qquad i = 1, ..., N. \tag{5.3}$$

By default, when calculating partial derivatives of the Lagrangian in (5.3), they are calculated in (q, v) variables.

We will assume that the Lagrangian (5.1) is constrained, which means that the second line in

Legendre Transformations (5.3) is not solvable for all velocities v_i, $i = 1,...,N$. It is solvable for only some of them or maybe even none. Since we can change the numeration of the variables, we can assume that the velocities for which we can solve are v_k, $k = 1,...,M < N$. So, we partially invert the second line in (5.3), getting:

$$\begin{aligned}v_k &= v_k(q, p, v_{M+1}, v_{M+2},..., v_N), \quad k = 1,...,M < N, \\ \varphi_s(q,p) &= 0, \quad s = M+1,...,N.\end{aligned} \quad (5.4)$$

It is possible that equations (5.3) cannot be solved for any velocities. This was the case in the example we covered in Chapter IV. If this happens, we will not have the first line in (5.4).

The expressions $\varphi_s(q,p) = 0$ in (5.4) are called primary constraints. By definition they will always appear for a constrained Lagrangian, and they will be independent of each other (meaning that removing any one of them will change the points (q, p) allowed by these constraints), and there must be $M - N$ of them. The starting equations (5.3) are independent, because on their left sides they have different symbols, and these symbols do not appear on the right side. For the same reason, the top N equations in (5.4) are independent. Since we assume (5.4) is equivalent to (5.3), the number of independent equations must be the same in both systems. If some of $\varphi_s(q,p) = 0$ were not independent, we would drop them. Therefore, there must be $M - N$ of them, because the number of independent equations must be the same for equivalent systems.

When we say that we solve the equations (5.3) for as many velocities as possible, we do not mean that we can do this explicitly. In specific examples, the formulas (5.3) may be so complicated that explicit solutions may be out of the question. So, what we mean is that we solve it "in principle." They could potentially be solved for, even if we cannot do it in practice.

The constraints $\varphi_s(q,p) = 0$ would not contain velocities anymore. If they contained any of the first M velocities, then we would use the top M equations in (5.4) to eliminate them. If there was a velocity higher than M in any constraint, then we would solve that constraint for this velocity, and include the solution among the top equations in (5.4).

Sometimes we prefer not to explicitly have the partial derivatives symbols as we analyze the Euler-Lagrange equations (5.2). Since we assume the Lagrangian (5.1) to be given, then also the partial derivatives of it are assumed to be known functions. We already have the symbols for partial derivatives with respect to velocities; they are given by the Legendre Transformations (5.3). Let us introduce new functions, F_i, as partial derivatives with respect to position, defined as:

$$F_i = F_i(q,v) = \left. \frac{\partial L}{\partial q_i} \right|_{(q,v)}, \qquad i = 1, ..., N. \tag{5.5}$$

By default, when calculating partial derivatives of the Lagrangian, the derivatives in (5.5) are calculated in (q,v) variables.

We do not intend to use F_i as new variables. Rather, they are functions that show up in the equations of motions. Possibly we could call them "forces" since, in the Euler-Lagrange equations, they appear as equal to time derivatives of canonical momenta. However, the reader should be aware that since our Lagrangian may not be "physically accepted Lagrangian", then also the time derivatives of our canonical momenta may not be what physicists would agree to call forces. Still, we may use the word "force" here; we just need to remember that it may not be a "physical" force.

Using dot over the symbol for the time derivative, the Legendre Transformations (5.3), and using F_i's from (5.5), we can rewrite the Euler-Lagrange equations (5.2) as:

$$\begin{aligned} \dot{p}_i &= F_i, \\ \dot{q}_i &= v_i, \end{aligned} \qquad i = 1, ..., N. \tag{5.6}$$

Let us combine this with the equations (5.4), getting:

$$\begin{aligned} \dot{p}_i &= F_i(q,v), \\ \dot{q}_i &= v_i, & i &= 1, ..., N. \\ v_k &= v_k(q, p, v_{M+1}, v_{M+2}, ..., v_N), & k &= 1, ..., M < N, \\ \varphi_s(q,p) &= 0, & s &= M+1, ..., N. \end{aligned} \tag{5.7}$$

It is convenient to split time derivatives of positions into the ones equal to velocities v_k, $k = 1,...,M < N$, and equal to v_m, $m = M+1,...,N$. We get:

$$\begin{aligned}
\dot{p}_i &= F_i(q,v), & & \\
\dot{q}_k &= v_k & k &= 1,...,M < N, \\
\dot{q}_m &= v_m & m &= M+1,...,N, \\
v_k &= v_k(q, p, v_{M+1}, v_{M+2},...,v_N), & k &= 1,...,M < N, \\
\varphi_s(q,p) &= 0, & s &= M+1,...,N.
\end{aligned} \qquad (5.8)$$

Then we replace velocities v_k, $k = 1,...,M$ in F_i, $i = 1,...,N$, by the formulas $v_k = v_k(q, p, v_{M+1}, v_{M+2},...,v_N)$, $k = 1,...,M < N$, from (5.8). We get:

$$\begin{aligned}
\dot{p}_i &= F_i(q, p, v_{M+1}, v_{M+2},...,v_N), & & \\
\dot{q}_k &= v_k(q, p, v_{M+1}, v_{M+2},...,v_N), & k &= 1,...,M < N, \\
\dot{q}_m &= v_m & m &= M+1,...,N, \\
v_k &= v_k(q, p, v_{M+1}, v_{M+2},...,v_N), & k &= 1,...,M < N, \\
\varphi_s(q,p) &= 0, & s &= M+1,...,N.
\end{aligned} \qquad (5.9)$$

It is important to realize that the system (5.9) is completely equivalent to the original Euler-Lagrange equations (5.2), just partially expressed by the momenta variables (5.3).

Now we are going through the completion process. The process is based on the fact that existing constraints may produce more constraints and/or express more velocities by positions and momenta. This is done by taking time derivatives of the existing constraints. Before doing that, we possibly need to modify how the constraints are written.

To understand the process let us look at a simple example. Say we have a constraint $x = 0$. Since constraints hold in time, the time derivatives of both sides must be equal. Therefore, we get $\dot{x} = 0$. Then we have $v = 0$. If the v is already expressed by positions and momenta, say $v = p$, then we get a new constraint $p = 0$. If the v was not expressed by positions and momenta yet, we simply get $v = 0$, which in a sense is an expression of v by positions and momenta, just a very simple one. It is also possible that we already have an equation $v = 0$ in our system, and then we get nothing from the time derivative of that constraint. We are happy with either of these results.

Imagine, that we express the constraint as $x^2 = 0$. It restricts the possible x's the same way as the constraint $x = 0$ does. But the time derivative is different. Using the chain rule, we get $2x\dot{x} = 0$. Then, since we already have $x = 0$, we simplify $2x\dot{x} = 0$ to $0 = 0$, which means there are no consequences of these constraints. However, this constraint has consequences, as we described in the previous paragraph. The conclusion is that a constraint must be written in a correct way. If we write it in an incorrect way, we may miss important consequences. So, what is the correct way? Without going into complicated explanations, we will just state that if a constraint is written in such a way that it is solved for one of the position or momenta variables, then it is written in a correct way, and the effect as the one above will be avoided. Once we have this form, we can move the one variable to the other side of the equation constraint, getting zero on the right side, which is a traditional way of writing constraints.

In practice, solving a constraint for a variable may be technically impossible, so as always, we really mean that we do it in principle, even if we cannot do it in practice.

So, say we have the system (5.9), in which the constraints are written in the way just described. To go through the completion process, we take time derivatives of constraints in (5.9). It will look like this:

$$\dot{\varphi}_s(q, p) = 0, \qquad s = M+1, \dots, N. \qquad (5.10)$$

The chain rule then gives:

$$\sum_{i=1}^{N} \frac{\partial \varphi_s}{\partial q_i} \cdot \dot{q}_i + \sum_{i=1}^{N} \frac{\partial \varphi_s}{\partial p_i} \cdot \dot{p}_i = 0, \qquad s = M+1, \dots, N. \qquad (5.11)$$

Then we replace $\dot{q}_i$ and $\dot{p}_i$ by the expressions from (5.9). We get:

$$\sum_{k=1}^{M} \frac{\partial \varphi_s}{\partial q_k} \cdot v_k \left(q, p, v_{M+1}, v_{M+2}, ..., v_N \right) + \sum_{k=M+1}^{N} \frac{\partial \varphi_s}{\partial q_k} \cdot v_k +$$
$$+ \sum_{i=1}^{N} \frac{\partial \varphi_s}{\partial p_i} \cdot F_i \left(q, p, v_{M+1}, v_{M+2}, ..., v_N \right) = 0, \qquad s = M+1, ..., N, \qquad (5.12)$$

and analyze the results for each s. The possibilities are as follows:

1. The expression (5.12) for an s simplifies to zero, possibly using the existing constraints. This means that this constraint has no consequences.

2. The expression (5.12) for an s can be solved for a velocity. Since the order of variables can be changed, we will assume this is the velocity v_{M+1}. It can then be expressed as

$$v_{M+1} = v_{M+1}\left(q, p, v_{M+2}, v_{M+3}, ..., v_N \right).$$

Then we look at other s's and check if they can be solved for the same v_{M+1}. If so, we get another equation that can give us another velocity, or if velocities are not present in it anymore, we may get another constraint containing just (q, p), so this is a new constraint, and we include it with constraints. Or maybe both expressions cancel out and produce no new constraints.
We repeat the process for all s's, possibly creating more expressions for velocities, and more constraints.

3. The expression for an s contains no velocities but is not simplifying to zero. This creates a new constraint.

4. We obtain a contradiction, for example $1 = 0$. This means that the Lagrangian with which we started is producing the Euler-Lagrange equations that are internally contradictory, and the entire Lagrangian should be discarded.

At the end we check the new and old constraints for independence. Independence means that removing any one constraint would change the points (q, p) allowed by these constraints.

In practice the independence of constraints can be studied in two ways. The first way is solving one of the constraints for one of the variables q_i or p_i, and then eliminating that variable in all other constraints using the result. Then we repeat the process for another constraint until we finish with all of them. The result will be the set of independent constraints and the set of identities of the kind $0 = 0$. The identities are then discarded. This process in practice may be very difficult or even impossible, so we describe it here as doable "in principle."

The second way is to calculate the matrix:

$$\begin{pmatrix} \frac{\partial \varphi_1}{\partial q_1} & \cdots & \frac{\partial \varphi_1}{\partial q_N} & \frac{\partial \varphi_1}{\partial p_1} & \cdots & \frac{\partial \varphi_1}{\partial p_N} \\ \vdots & & \vdots & & & \\ \frac{\partial \varphi_S}{\partial q_1} & \cdots & \frac{\partial \varphi_S}{\partial q_N} & \frac{\partial \varphi_S}{\partial p_1} & & \frac{\partial \varphi_S}{\partial p_N} \end{pmatrix}$$

If the rank of this matrix is equal to the number of constraints, then the constraints are independent. Without going into detail, if the rank of this matrix is smaller than the number of constraints, then the matrix may be used to identify the dependent constraints. As in the earlier case, this process, described here in principle, in practice may be very difficult or even impossible.

Then we include all independent secondary constraints into the existing equations (5.9). We also express velocities obtained in point 2 by (q, p). We get:

$$\begin{aligned} \dot{p}_i &= F_i\left(q, p, v_{M_1+1}, v_{M_1+2}, \ldots, v_N\right), & & \\ \dot{q}_k &= v_k\left(q, p, v_{M_1+1}, v_{M_1+2}, \ldots, v_N\right), & k &= 1, \ldots, M_1 < N, \\ \dot{q}_m &= v_m & m &= M_1+1, \ldots, N, \\ v_k &= v_k\left(q, p, v_{M_1+1}, v_{M_1+2}, \ldots, v_N\right), & k &= 1, \ldots, M_1 < N, \\ \varphi_s(q, p) &= 0, & s &= M_2+1, \ldots, N. \end{aligned} \quad (5.13)$$

The equations (5.13) look the same as (5.9); just the number of velocities that are not expressed by (q, p) is possibly smaller, and the number of constraints in the last line may be larger.

Now, we repeat the process of taking the time derivatives, but only of the new constraints, since the time derivatives of old constraints are simplified to zero. We repeat this process until time derivatives of all constraints are simplified to zero, possibly using the constraints obtained earlier in the simplification process. The process has a finite number of steps since the number of velocities is finite and the number of possible independent constraints is finite. Since the process stops when expressions for new velocities and new independent constants stop appearing, it will end in a finite number of steps.

Now we observe that when this process is finished, all velocities will be expressed by (q, p). This is because if they were not, these velocities would be completely free as functions of time when solving the system (5.13). We could choose them freely, and then solve the system for other variables. However, this freedom is exactly what we earlier defined as a gauge. We agreed that our Lagrangian is regular, and this, by the definition, means that there is no gauge.

In other words, if a gauge exists at the end of this process, we simply discard this Lagrangian as not interesting to us. Actually, it is very interesting, but gauge is a separate and such a large topic that we simply decided not to include it in this book.

Concluding, for a regular constrained Lagrangian, at the end of this process we get the equations that look like:

$$\dot{p}_i = F_i(q, p),$$
$$\dot{q}_i = v_i(q, p),$$
$$v_i = v_i(q, p), \qquad i = 1, ..., N, \qquad (5.14)$$
$$\varphi_s(q, p) = 0, \qquad s = 1, ..., S.$$

Finally, when using the Hamilton's equations, we do not use the velocities anymore. So, we drop the velocities from the equations (5.14). Obviously, we can keep a record of how velocities are related to positions and momenta. However, it is not really needed, since we can observe that the record of what velocities are equal to, still is in the formulas for $\dot{q}_i$, $i = 1, ..., N$.

So, we finally get the equations:

$$\begin{aligned}&\dot{p}_i = F_i(q,p), \\ &\dot{q}_i = v_i(q,p), \qquad\qquad i=1,...,N, \\ &\varphi_s(q,p) = 0, \qquad\qquad s=1,...,S.\end{aligned} \qquad (5.15)$$

The equations (5.15) are the Hamilton's equations for the regular constrained Lagrangian (5.1). It is important to realize that the Hamilton's equations (5.15) are completely equivalent to the Euler-Lagrange equations (5.2).

Notice that because we were removing the dependent constraints at each step of the process, the constraints appearing in (5.15) are independent, in the sense that removing any one of them would change the points (q,p) that satisfy them.

It is also important to observe that the expressions $F_i(q,p)$ and $v_i(q,p)$ appearing in (5.15) will be, in principle, fully calculated as a result of the completion process from the initial Euler-Lagrange equations (5.2) and the Legendre Transformations (5.3).

Also, let us point out that the Hamilton's equations (5.15) were obtained without calculating the Hamiltonian.

2. Regular Constrained Lagrangian, the Newtonian Equations

The Newtonian equations (1.4) are often important to obtain, since they use the variables (q,v), which are directly related to physical observations.

Below, we describe a general procedure of obtaining the Newtonian equations from a constrained Lagrangian. The procedure shows that, in principle, Newtonian equations always exist for regular constrained Lagrangian. However, in specific cases, the procedure may be sometimes simplified compared to the general case we show here.

We can start by going back to equations (5.14):

$$\dot{p}_i = F_i(q,p),$$
$$\dot{q}_i = v_i(q,p),$$
$$v_i = v_i(q,p), \qquad\qquad i = 1,\ldots,N,$$
$$\varphi_s(q,p) = 0, \qquad\qquad s = 1,\ldots,S.$$

Taking the time derivative of both sides of line 3 of the above formulas and using the chain rule, we get:

$$\begin{aligned}
\dot{p}_i &= F_i(q,p), \\
\dot{q}_i &= v_i(q,p), \\
\dot{v}_i &= \sum_{k=1}^{N} \frac{\partial v_i(q,p)}{\partial q_k} \cdot \dot{q}_k + \sum_{k=1}^{N} \frac{\partial v_i(q,p)}{\partial p_k} \cdot \dot{p}_k, \\
v_i &= v_i(q,p), \qquad\qquad i = 1,\ldots,N, \\
\varphi_s(q,p) &= 0, \qquad\qquad s = 1,\ldots,S.
\end{aligned} \qquad (5.16)$$

Now we replace the time derivatives on the right side of expressions in line 3 with the expression for these derivatives appearing in the first two lines, getting:

$$\begin{aligned}
\dot{p}_i &= F_i(q,p), \\
\dot{q}_i &= v_i(q,p), \\
\dot{v}_i &= \sum_{k=1}^{N} \frac{\partial v_i(q,p)}{\partial q_k} \cdot v_i(q,p) + \sum_{k=1}^{N} \frac{\partial v_i(q,p)}{\partial p_k} \cdot F_i(q,p), \\
v_i &= v_i(q,p), \qquad\qquad i = 1,\ldots,N, \\
\varphi_s(q,p) &= 0, \qquad\qquad s = 1,\ldots,S.
\end{aligned} \qquad (5.17)$$

We will not use lines 1 and 4 anymore, so we drop them, getting:

$$\begin{aligned}
\dot{q}_i &= v_i(q,p), \\
\dot{v}_i &= \sum_{k=1}^{N} \frac{\partial v_i(q,p)}{\partial q_k} \cdot v_i(q,p) + \sum_{k=1}^{N} \frac{\partial v_i(q,p)}{\partial p_k} \cdot F_i(q,p), \qquad i = 1,\ldots,N, \\
\varphi_s(q,p) &= 0, \qquad\qquad s = 1,\ldots,S.
\end{aligned} \qquad (5.18)$$

Now we change all variables from (q,p) to (q,v). We get:

$$\dot{q}_i = v_i(q,p)\Big|_{p_i=\frac{\partial L}{\partial v_i}}, \qquad (5.19a)$$

$$\dot{v}_i = \sum_{k=1}^{N} \frac{\partial v_i(q,p)}{\partial q_k}\Big|_{p_i=\frac{\partial L}{\partial v_i}} \cdot v_i(q,p)\Big|_{p_i=\frac{\partial L}{\partial v_i}} +$$

$$+\sum_{k=1}^{N} \frac{\partial v_i(q,p)}{\partial p_k}\Big|_{p_i=\frac{\partial L}{\partial v_i}} \cdot F_i(q,p)\Big|_{p_i=\frac{\partial L}{\partial v_i}}, \qquad i=1,...,N, \qquad (5.19b)$$

$$\varphi_s(q,p)\Big|_{p_i=\frac{\partial L}{\partial v_i}} = 0, \qquad s=1,...,S.$$

Let us stress that in (5.19) the partial derivatives of velocities are calculated and expressed using the variables (q,p), while the partial derivatives of the Lagrangian are calculated and expressed using the variables (q,v), as agreed earlier.

In (5.19) the $v_i(q,p)\Big|_{p_i=\frac{\partial L}{\partial v_i}}$ are velocities expressed in variables (q,v). Velocity expressed by (q,v) must be just v. So, we have:

$$v_i(q,p)\Big|_{p_i=\frac{\partial L}{\partial v_i}} = v_i$$

Then $F_i(q,p)\Big|_{p_i=\frac{\partial L}{\partial v_i}}$ are forces expressed in variables (q,v). Then, we go back to equations (5.5) and we get:

$$F_i(q,p)\bigg|_{p_i=\frac{\partial L}{\partial v_i}} = F_i(q,v) = \frac{\partial L}{\partial q_i}\bigg|_{(q,v)},$$

Placing the last two equations in (5.19), we get:

$$\dot{q}_i = v_i, \tag{5.20a}$$

$$\dot{v}_i = \sum_{k=1}^{N} \frac{\partial v_i(q,p)}{\partial q_k}\bigg|_{p_i=\frac{\partial L}{\partial v_i}} \cdot v_i + \sum_{k=1}^{N} \frac{\partial v_i(q,p)}{\partial p_k}\bigg|_{p_i=\frac{\partial L}{\partial v_i}} \cdot \frac{\partial L}{\partial q_i}\bigg|_{(q,v)},$$

$$\varphi_s(q,p)\bigg|_{p_i=\frac{\partial L}{\partial v_i}} = 0, \qquad i=1,\ldots,N, \qquad s=1,\ldots,S. \tag{5.20b}$$

Finally, we switch the order of the first two lines just to get the order that we prefer. We get:

$$\dot{v}_i = \sum_{k=1}^{N} \frac{\partial v_i(q,p)}{\partial q_k}\bigg|_{p_i=\frac{\partial L}{\partial v_i}} \cdot v_i + \sum_{k=1}^{N} \frac{\partial v_i(q,p)}{\partial p_k}\bigg|_{p_i=\frac{\partial L}{\partial v_i}} \cdot \frac{\partial L}{\partial q_i}\bigg|_{(q,v)},$$

$$\dot{q}_i = v_i, \tag{5.21}$$

$$\varphi_s(q,p)\bigg|_{p_i=\frac{\partial L}{\partial v_i}} = 0, \qquad i=1,\ldots,N, \qquad s=1,\ldots,S.$$

The equations (5.21) are the Newtonian equations for the constrained Lagrangian (5.1). Notice that, in general, we cannot get the partial derivatives of velocities in (5.21) without first going through the completion process ending with the equations (5.14).

Also, it is very important to realize that in the equations (5.21), the partial derivatives of velocities must be calculated using the variables (q,p), while the partial derivatives of the Lagrangian must be calculated using the variables (q,v).

Observe that the constraints showing up in the Newtonian equations (5.21) are the same constraints that show up in the Hamilton's equations. They are just expressed in the (q,v) variables.

Finally, we introduce new symbols that we will call accelerations:

$$a_i = \sum_{k=1}^{N} \left.\frac{\partial v_i(q,p)}{\partial q_k}\right|_{p_i=\frac{\partial L}{\partial v_i}} \cdot v_i + \sum_{k=1}^{N} \left.\frac{\partial v_i(q,p)}{\partial p_k}\right|_{p_i=\frac{\partial L}{\partial v_i}} \cdot \left.\frac{\partial L}{\partial q_i}\right|_{(q,v)}, \qquad i=1,\ldots,N, \qquad (5.22)$$

and we rewrite the constraints from the last line of (5.21) in terms of positions and velocities as:

$$\vartheta_s(q,v) = \left.\varphi_s(q,p)\right|_{p_i=\frac{\partial L}{\partial v_i}} = 0, \qquad i=1,\ldots,N, \quad s=1,\ldots,S. \qquad (5.23)$$

Using (5.22) and (5.23), the equations (5.21) can be rewritten as:

$$\begin{aligned}
\dot{v}_i &= a_i(q,v), \\
\dot{q}_i &= v_i, \\
\vartheta_s(q,v) &= 0, \qquad i=1,\ldots,N, \quad s=1,\ldots,S.
\end{aligned} \qquad (5.24)$$

The equations (5.24) show a general form of Newtonian equations of motion in the case of regular constrained Lagrangian.

3. A Comment

Notice that Hamilton's equations of motion, as well as the Newtonian equations of motion, were obtained from the Euler-Lagrange equations of motion without the need to introduce the Hamiltonian.

CHAPTER VI

GENERAL DESCRIPTION OF CONSTRAINED FUNCTIONS. EQUATIONS OF MOTION FOR CONSTRAINED FUNCTIONS

When dealing with constraints, we find ourselves in the situation where some of the variables can be expressed by others using the constraints and substituted into functions. It means that, in general, the same function of the same variables can be expressed in many ways, in a rather arbitrary manner. In this chapter, we will study the situation in detail, giving an organized description of how such substitutions are made. We will also study how such substitutions affect the partial derivatives of functions.

In all considerations in this Chapter, we will assume that all functions are smooth up to some order, meaning that their partial derivatives exist up to some, possibly infinite, order. We will not go into details here since such considerations would be too long.

1. An Example of Functions Equal on Constraints

We will start by addressing an example of a situation when we have a function of many variables, but the variables are not independent; some variables in the function are related to other variables in it. This happens in many instances and may be described by using the words "constraints," "substitution," "replacement," or a "change of variables".

An example:

Say we have a function of three variables:

$$f(x, y, z) = x^2 y z^3 - 7yz + xz\sin(2xy + z) + z. \tag{6.1}$$

We have no special reason for choosing this function for our consideration. We just need a function with multiple appearances of each variable.

Assume also that we have a constraint that relates x, y and z. We want to avoid complicated calculations to possibly cloud the main point, so for our example we choose a rather simple relation:

$$xyz = 5$$

or, equivalently

$$x = \frac{5}{yz}, \ y = \frac{5}{xz}, \ \text{or} \ z = \frac{5}{xy}.$$

A constraint of this kind defines a 2-dimensional subspace of the original 3-dimensional (x, y, z) space.

Let us now say that we want to restrict our original function $f(x, y, z)$ only to the subspace defined by that constraint and write some expressions that will describe the restricted function. How can we do it? The answer is it can be done in infinitely many ways. Let us look at some of the possibilities:

a) We can replace x by $\dfrac{5}{yz}$ in all places where x appears, getting:

$$f(x, y, z) = \frac{25z}{y} - 7yz + \frac{5}{y}\sin\left(\frac{25}{z} + z\right) + z \tag{6.2}$$

Is the function given by (6.2) identical to the function in (6.1)? Not on the entire 3-dimensional space (x, y, z). However, if we restrict ourselves to using only the points that satisfy the constraint $xyz = 5$, then yes, they are equal. In this sense one can be replaced by the other.

Still, we have to be careful when calculating partial derivatives since the partial derivatives will be different, even for the points that satisfy the constraints. We will discuss the issue in more detail in the next section.

Similarly, we can remove y instead of x, or we can remove z instead of x. Because it is so similar to the process we just described, we are not going to show it here.

b) We can do a partial replacement of a variable. For example:

$$f(x,y,z) = x^2 y z^3 - 7yz + xz \sin(2xy+z) + z$$

replaced by

$$f(x,y,z) = \frac{25z}{y} - 7yz + xz \sin(2xy+z) + z$$

In the above, we replaced x by $\frac{5}{yz}$ in the first term only. The resulting function is equal to the original function, as long as we only consider x's, y's, and z's that satisfy the constraint $xyz = 5$.

c) We can add a function that is equal to 0 for points satisfying the constraint $xyz = 5$. For example:

$$f(x,y,z) = x^2 y z^3 - 7yz + xz \sin(2xy+z) + z + x^2(xyz - 5)$$

is equal to the original $f(x,y,z) = x^2 y z^3 - 7yz + xz \sin(2xy+z) + z$ for the points satisfying the constraint $xyz = 5$.

What can be concluded from the above examples is that there is an infinite variety of possible modifications of a given function into another function that is equal to the original function for the subspace described by given constraints. So, the obvious question is, "How can we characterize all such functions?" It turns out that the answer is quite simple.

2. Generalized Substitutions for a Single Constraint

We will start with a system with just one constraint. It means that we assume that we have a single constraint, and we will write it in the form $\varphi(x_1, x_2, x_3, ..., x_n) = 0$. We will assume that

$\varphi(x_1, x_2, x_3, ..., x_n)$ and all other functions in the remainder of this chapter are smooth up to some order.

Also, assume we have two functions, denoted by $f(x_1, x_2, x_3, ..., x_n)$ and $g(x_1, x_2, x_3, ..., x_n)$, and that these two functions are equal for all $(x_1, x_2, x_3, ..., x_n)$ satisfying the constraint $\varphi(x_1, x_2, x_3, ..., x_n) = 0$.

Then define another function:

$$u(x_1, x_2, x_3, ..., x_n) = \frac{g(x_1, x_2, x_3, ..., x_n) - f(x_1, x_2, x_3, ..., x_n)}{\varphi(x_1, x_2, x_3, ..., x_n)} \quad \text{if} \quad \varphi(x_1, x_2, x_3, ..., x_n) \neq 0 \qquad (6.3)$$

For the points when $\varphi(x_1, x_2, x_3, ..., x_n) = 0$, we define $u(x_1, x_2, x_3, ..., x_n)$ as continuous extension of the $u(x_1, x_2, x_3, ..., x_n)$ defined by (6.3). Without going to details, such extension will always exist, if we properly re-define the constraint function $\varphi(x_1, x_2, x_3, ..., x_n)$, so the points allowed by the re-defined constraint will be the same, and the $u(x_1, x_2, x_3, ..., x_n)$, defined by (6.3), will have only removable discontinuities.

Solving the above for $g(x_1, x_2, x_3, ..., x_n)$, we get a very important result:

$$g(x_1, x_2, x_3, ..., x_n) = f(x_1, x_2, x_3, ..., x_n) + u(x_1, x_2, x_3, ..., x_n)\varphi(x_1, x_2, x_3, ..., x_n). \qquad (6.4)$$

This result means that starting with any function $f(x_1, x_2, x_3, ..., x_n)$ we can obtain any other function $g(x_1, x_2, x_3, ..., x_n)$ if the two functions are equal for the points given by a constraint $\varphi(x_1, x_2, x_3, ..., x_n) = 0$, by adding the constraint function $\varphi(x_1, x_2, x_3, ..., x_n)$ multiplied by another function $u(x_1, x_2, x_3, ..., x_n)$.

So, we can see that to get all functions that are equal to a given function for points satisfying the constraint, we can start with a given function, and we can get any other by adding the constraint formula multiplied by other functions. Results like that allow us to better understand how all functions equal on a constraint look like.

Notice that substitutions, defined as using a constraint to replace some variables with others, produce functions equal to the original functions on the constraint. Therefore, if we have $g(x_1, x_2, x_3, ..., x_n)$ obtained from $f(x_1, x_2, x_3, ..., x_n)$ by a substitution using the constraint, we also have

$$g(x_1, x_2, x_3, ..., x_n) = f(x_1, x_2, x_3, ..., x_n) + u(x_1, x_2, x_3, ..., x_n)\varphi(x_1, x_2, x_3, ..., x_n). \tag{6.5}$$

Therefore, we will call the expression (6.5) a generalized substitution. This definition expresses the fact that using a constraint to make a substitution in a function is always equivalent to adding the constraint multiplied by a suitable function to the original function.

3. Generalized Substitutions for Multiple Constraints

The question we are going to look at now is, "Will a result analogous to the one obtained in Section 2 above still hold when instead of one constraint, we will have multiple constraints?"

Assume we have several constraints:

$$\varphi_i(x_1, x_2, ..., x_N) = 0, \qquad i = 1, ..., M, \qquad M < N. \tag{6.6}$$

The points $(x_1, x_2, ..., x_n)$ satisfying the constraints are, by definition, such points $(x_1, x_2, ..., x_n)$ that satisfy all the constraints simultaneously.

We will not assume any kind of constraint independence of each other at this point. (In general, independence would mean that removing any one constraint from the set would change the points $(x_1, x_2, ..., x_n)$ that satisfy this set of constraints.)

Now take any functions $u_i(x_1, x_2, ..., x_N)$, $i = 1, ..., M$, that are as smooth as $f(x_1, x_2, ..., x_n)$.

Then, since $\sum_{i=1}^{M} u_i(x_1, x_2, x_3, ..., x_N) \varphi_i(x_1, x_2, x_3, ..., x_N)$ is equal to 0 for all points satisfying the constraints, also the function $g(x_1, x_2, x_3, ..., x_n)$ defined by:

$$g(x_1, x_2, ..., x_n) = f(x_1, x_2, ..., x_N) + \sum_{i=1}^{M} u_i(x_1, x_2, ..., x_N) \varphi_i(x_1, x_2, ..., x_N) \qquad (6.7)$$

will be equal to $f(x_1, x_2, x_3, ..., x_n)$, for all points satisfying the constraints. In other words, adding $\sum_{i=1}^{M} u_i(x_1, x_2, x_3, ..., x_N) \varphi_i(x_1, x_2, x_3, ..., x_N)$ to a function, gives a function equal to the original function on the constraints.

The question is, "Is this addition giving us all functions equal on constraints?" In other words, if we have two functions, $f(x_1, x_2, ..., x_N)$ and $g(x_1, x_2, ..., x_N)$, such that they are equal for each point $(x_1, x_2, ..., x_n)$ satisfying the constraints, would such functions $u_i(x_1, x_2, ..., x_N)$, $i = 1, ..., M$ exists that:

$$g(x_1, x_2, ..., x_N) = f(x_1, x_2, ..., x_N) + \sum_{i=1}^{M} u_i(x_1, x_2, ..., x_N) \varphi_i(x_1, x_2, ..., x_N) \qquad ?$$

To construct such $u_i(x_1, x_2, ..., x_N)$, let us look at a single, specific point that does not satisfy the constraints. Then, by definition, there must be at least one constraint such that for this one specific point the constraint will not be equal to zero. Because of smoothness, there is a neighborhood of the point on which this constraint is not zero. Since we can re-numerate the constraints as we wish, we can assume that it is the last constraint. So, say $\varphi_M(x_1, x_2, ..., x_N) \neq 0$ for the neighborhood.

Then, for the neighborhood, define the functions $u_i(x_1, x_2, ..., x_N)$, $i = 1, ..., M-1$ in any way we want. We make them smooth to the extent that we need.

When this is done, we define $u_M(x_1, x_2, ..., x_N)$ for the neighborhood as:

$$u_M(x_1, x_2, ..., x_N) = \frac{g(x_1, x_2, ..., x_N) - f(x_1, x_2, ..., x_N) - \sum_{i=1}^{M-1} u_i(x_1, x_2, ..., x_N) \varphi_i(x_1, x_2, ..., x_N)}{\varphi_M(x_1, x_2, ..., x_N)} \tag{6.8}$$

For points for which $\varphi_M(x_1, x_2, ..., x_N) = 0$, we define $u_M(x_1, x_2, ..., x_N)$ by continuous extension of (6.8). This can be done if the constraint functions and other $u_i(x_1, x_2, ..., x_N)$ are chosen properly.

Then, for the neighborhood we have:

$$f(x_1, x_2, ..., x_N) + \sum_{i=1}^{M} u_i(x_1, x_2, ..., x_N) \varphi_i(x_1, x_2, ..., x_N) =$$

$$= f(x_1, x_2, ..., x_N) + \sum_{i=1}^{M-1} u_i(x_1, x_2, ..., x_N) \varphi_i(x_1, x_2, ..., x_N) + u_M(x_1, x_2, ..., x_N) \varphi_M(x_1, x_2, ..., x_N) =$$

$$= f(x_1, x_2, ..., x_N) + \sum_{i=1}^{M-1} u_i(x_1, x_2, ..., x_N) \varphi_i(x_1, x_2, ..., x_N) +$$

$$+ \frac{g(x_1, x_2, ..., x_N) - f(x_1, x_2, ..., x_N) - \sum_{i=1}^{M-1} u_i(x_1, x_2, ..., x_N) \varphi_i(x_1, x_2, ..., x_N)}{\varphi_M(x_1, x_2, ..., x_N)} \cdot \varphi_M(x_1, x_2, ..., x_N) =$$

$$= f(x_1, x_2, ..., x_N) + \sum_{i=1}^{M-1} u_i(x_1, x_2, ..., x_N) \varphi_i(x_1, x_2, ..., x_N) +$$

$$+ g(x_1, x_2, ..., x_N) - f(x_1, x_2, ..., x_N) - \sum_{i=1}^{M-1} u_i(x_1, x_2, ..., x_N) \varphi_i(x_1, x_2, ..., x_N) =$$

$$= g(x_1, x_2, ..., x_N)$$

So, for the neighborhood we get:

$$g(x_1, x_2, ..., x_N) = f(x_1, x_2, ..., x_N) + \sum_{i=1}^{M} u_i(x_1, x_2, ..., x_N) \varphi_i(x_1, x_2, ..., x_N) \tag{6.9}$$

Concluding, for any two functions $f(x_1, x_2, ..., x_N)$ and $g(x_1, x_2, ..., x_N)$ that are equal on constraints $\varphi_i(x_1, x_2, ..., x_N) = 0$, $i = 1, ..., M$, $M < N$, there (locally) exist such functions $u_i(x_1, x_2, ..., x_N)$, $i = 1, ..., M$ that:

$$g(x_1, x_2, ..., x_N) = f(x_1, x_2, ..., x_N) + \sum_{i=1}^{M} u_i(x_1, x_2, ..., x_N) \varphi_i(x_1, x_2, ..., x_N). \tag{6.10}$$

As an observation, notice that what we imprecisely call substitutions, is re-writing a function using constraints. Therefore any $g(x_1, x_2, ..., x_N)$ obtained from $f(x_1, x_2, ..., x_N)$ by a substitution is equal to $f(x_1, x_2, ..., x_N)$ on constraints, so it satisfies:

$$g(x_1, x_2, ..., x_N) = f(x_1, x_2, ..., x_N) + \sum_{i=1}^{M} u_i(x_1, x_2, ..., x_N) \varphi_i(x_1, x_2, ..., x_N) \tag{6.11}$$

with some $u_i(x_1, x_2, ..., x_N)$, $i = 1, ..., M$.

So, if we want to study the general properties of substitutions, we can use the expression (6.11).

Finally, since we were choosing quite freely the first $M-1$ functions $u_i(x_1, x_2, ..., x_N)$, $i = 1, ..., M-1$, and only calculating $u_M(x_1, x_2, ..., x_N)$ later, the choice of the $u_i(x_1, x_2, ..., x_N)$, $i = 1, ..., M$, for a given $f(x_1, x_2, ..., x_N)$ and $g(x_1, x_2, ..., x_N)$ is not unique. There are infinitely many choices for such $u_i(x_1, x_2, ..., x_N)$.

4. An Example of Derivatives for Functions Equal on Constraints

The obvious question is, "If two functions are equal on constraints, would their partial derivatives also be equal?" We will start with a simple example with just two variables and one constraint. Let us have:

$f(x, y) = x$ and the constraint be $\varphi(x, y) = xy - 1 = 0$.

Consider $u(x,y) = -\dfrac{1}{y}$, and define:

$$g(x,y) = f(x,y) + u(x,y)\varphi(x,y)$$

We already know that $f(x,y)$ and $g(x,y)$ are equal on the constraint $\varphi(x,y)=0$. A simple calculation gives:

$$g(x,y) = f(x,y) + u(x,y)\varphi(x,y) = x + \left(-\frac{1}{y}\right)\cdot(xy-1) = x - \frac{xy}{y} + \frac{1}{y} =$$

$$= x - x + \frac{1}{y} = \frac{1}{y}$$

So, the functions

$$f(x,y) = x$$

$$g(x,y) = \frac{1}{y}$$

are equal on the constraint $\varphi(x,y) = xy-1 = 0$.

Are their partial derivatives also equal on the same constraint? Well, to answer this question, we need to specify what we mean by partial derivatives. When calculating partial derivatives, we specify not only the variable with respect to which we take derivative, but also the variables that are kept constant in this process.

Here, when we take derivative with respect to x, we assume that y is kept constant. When we take derivative with respect to y, we assume that x is kept constant. Notice that this assumption is not in agreement with the constraint. If we look at the constraint, then modifying x in the process of calculating the derivative, the constraint would enforce that y was not constant.

However, in typical calculations of Classical Mechanics, when we use partial derivatives, we ignore the constraints. When we calculate the partial derivatives in the Euler-Lagrange equations, we ignore the constraints. When we calculate the partial derivatives in Hamilton's equations, we ignore the constraints. We ignore the constraints because the very definitions of the Euler-Lagrange and Hamilton's equations for constrained systems tell us to ignore the constraints when calculating the partial derivatives.

Similarly, when we calculate the Poisson Brackets and the Dirac's Brackets, we ignore the constraints when calculating partial derivatives. Again, ignoring constraints is included in the definition of these brackets.

So, the definitions of basic objects in Classical Mechanics ignore constraints. There are at least two very good reasons to ignore them. First, we often calculate the derivatives first, and only later we can see what the constraints are. Second, if we have more than two variables in a constraint, then it is not clear at all how the other variables would change when we change one variable in the process of taking a partial derivative. It would be quite a mess to try to look at all possibilities.

So, the only reasonable choice is to always ignore constraints when calculating partial derivatives, and then study the consequences of such an approach.

Using this approach we get:

$$\frac{\partial f(x,y)}{\partial x} = 1$$

$$\frac{\partial g(x,y)}{\partial x} = 0$$

These two results are obviously not equal, even on constraints.

So, a simple observation is that even if two functions are equal on constraints, their partial derivatives do not have to be, even when restricted to the constraints.

5. Derivatives of Functions Equal on Constraints

We already know from (6.11), that if we start with a function $f(x_1, x_2, ..., x_N)$ then the general expression for all functions equal to it on constraints is:

$$f(x_1, x_2, ..., x_N) + \sum_{i=1}^{M} u_i(x_1, x_2, ..., x_N) \varphi_i(x_1, x_2, ..., x_N) \tag{6.12}$$

Taking a partial derivative of it while, as described earlier, ignoring possible relation between variables resulting from constraints, we get:

$$\frac{\partial f(x_1, x_2, ..., x_N)}{\partial x_k} + \frac{\partial}{\partial x_k}\left[\sum_{i=1}^{M} u_i(x_1, x_2, ..., x_N) \varphi_i(x_1, x_2, ..., x_N)\right] =$$

$$= \frac{\partial f(x_1, x_2, ..., x_N)}{\partial x_k} + \sum_{i=1}^{M} \frac{\partial}{\partial x_k} u_i(x_1, x_2, ..., x_N) \cdot \varphi_i(x_1, x_2, ..., x_N) +$$

$$+ \sum_{i=1}^{M} u_i(x_1, x_2, ..., x_N) \frac{\partial}{\partial x_k} \varphi_i(x_1, x_2, ..., x_N) =$$

$$= \frac{\partial f(x_1, x_2, ..., x_N)}{\partial x_k} + \sum_{i=1}^{M} \frac{\partial}{\partial x_k} u_i(x_1, x_2, ..., x_N) \cdot 0 +$$

$$+ \sum_{i=1}^{M} u_i(x_1, x_2, ..., x_N) \frac{\partial}{\partial x_k} \varphi_i(x_1, x_2, ..., x_N) =$$

$$= \frac{\partial f(x_1, x_2, ..., x_N)}{\partial x_k} + \sum_{i=1}^{M} u_i(x_1, x_2, ..., x_N) \frac{\partial}{\partial x_k} \varphi_i(x_1, x_2, ..., x_N).$$

So, we get:

$$\frac{\partial f(x_1, x_2, ..., x_N)}{\partial x_k} + \sum_{i=1}^{M} u_i(x_1, x_2, ..., x_N) \frac{\partial}{\partial x_k} \varphi_i(x_1, x_2, ..., x_N) \tag{6.13}$$

However, if we are only interested in the derivative on the constraints then, as described in (6.11), we can add $\sum_{i=1}^{M} w_i(x_1, x_2, ..., x_N) \varphi_i(x_1, x_2, ..., x_N)$ to the result, getting:

$$\frac{\partial f(x_1, x_2, ..., x_N)}{\partial x_k} + \sum_{i=1}^{M} u_i(x_1, x_2, ..., x_N) \frac{\partial}{\partial x_k} \varphi_i(x_1, x_2, ..., x_N) +$$
$$+ \sum_{i=1}^{M} w_i(x_1, x_2, ..., x_N) \varphi_i(x_1, x_2, ..., x_N) \tag{6.14}$$

So, we have quite a lot of freedom in the final result.

6. Effects of Constraints on the Equations of Motion

We will now shortly discuss the effects of the generalized substitution on the equations of motion. Please notice that when we look at typical equations of motion, in particular the Newtonian equations of motion (1.3), the Euler-Lagrange equations of motion (2.3), or some of the Hamilton's equations of motion, namely (3.5), then every equation in these systems has one of the two forms:

$$\frac{d}{dt} f(x_1, x_2, ..., x_N) = F(x_1, x_2, ..., x_N) \quad \text{(a differential equation of motion)} \tag{6.15}$$

or

$$\varphi(x_1, x_2, ..., x_N) = 0 \quad \text{(a constraint)} \tag{6.16}$$

In the above, the variables $(x_1, x_2, ..., x_N)$ represent all possible variables that we use to describe the system, for example, positions, velocities, momenta, or some other variables. Therefore the N is not the space dimension of the system.

In the above, we assume that $F(x_1, x_2, ..., x_N)$ is an explicitly given function of $(x_1, x_2, ..., x_N)$, and is not given by partial derivatives, like when using the Poisson Brackets. The case of Hamilton's equations of motion given by partial derivatives of a Hamiltonian (3.14), or the Poisson Brackets of a Hamiltonian (3.21) or (3.22), are considered separately in chapter VIII.

Let us start with the left side of the equation of motion (6.15). Say that we replace

$$f(x_1, x_2, ..., x_N) \tag{6.17}$$

by

$$f(x_1, x_2, ..., x_N) + \sum_{i=1}^{M} u_i(x_1, x_2, ..., x_N) \varphi_i(x_1, x_2, ..., x_N),$$ (6.18)

where $\varphi_i(x_1, x_2, ..., x_N)$ are the constraints in the system, and $u_i(x_1, x_2, ..., x_N)$ are any smooth enough functions.

The left side of (6.15) is the time derivative, so let us look at the time derivative of (6.18). We get:

$$\frac{d}{dt}\left[f(x_1, x_2, ..., x_N) + \sum_{i=1}^{M} u_i(x_1, x_2, ..., x_N) \varphi_i(x_1, x_2, ..., x_N)\right] =$$

$$= \frac{d}{dt} f(x_1, x_2, ..., x_N) + \frac{d}{dt}\left[\sum_{i=1}^{M} u_i(x_1, x_2, ..., x_N) \varphi_i(x_1, x_2, ..., x_N)\right] =$$

$$= \frac{d}{dt} f(x_1, x_2, ..., x_N) + \sum_{i=1}^{M} \frac{d}{dt}\left[u_i(x_1, x_2, ..., x_N) \cdot \varphi_i(x_1, x_2, ..., x_N)\right] =$$

$$= \frac{d}{dt} f(x_1, x_2, ..., x_N) + \sum_{i=1}^{M} \frac{d}{dt} u_i(x_1, x_2, ..., x_N) \cdot \varphi_i(x_1, x_2, ..., x_N) +$$

$$+ \sum_{i=1}^{M} u_i(x_1, x_2, ..., x_N) \cdot \frac{d}{dt} \varphi_i(x_1, x_2, ..., x_N) = \frac{d}{dt} f(x_1, x_2, ..., x_N) +$$

$$+ \sum_{i=1}^{M} \frac{d}{dt} u_i(x_1, x_2, ..., x_N) \cdot 0 + \sum_{i=1}^{M} u_i(x_1, x_2, ..., x_N) \cdot 0 =$$

$$= \frac{d}{dt} f(x_1, x_2, ..., x_N)$$

The line before the last one resulted from the fact that any constraint and its time derivative are equal to zero for all solutions of the equations of motion since the solutions must satisfy the constraints.

Concluding, we have:

$$\frac{d}{dt}\left[f(x_1, x_2, ..., x_N) + \sum_{i=1}^{M} u_i(x_1, x_2, ..., x_N) \varphi_i(x_1, x_2, ..., x_N)\right] = \frac{d}{dt} f(x_1, x_2, ..., x_N)$$ (6.19)

It means that any substitution using constraints on the left side of a differential equation of motion

(6.15), will give a correct equation of motion. It means that we can use substitutions when simplifying the left side of the equations of the differential equations of motion.

Now let us look at the right side of the equation of motion (6.15). Say that we replace

$$F(x_1, x_2, ..., x_N) \tag{6.20}$$

by

$$F(x_1, x_2, ..., x_N) + \sum_{i=1}^{M} u_i(x_1, x_2, ..., x_N) \varphi_i(x_1, x_2, ..., x_N), \tag{6.21}$$

where $\varphi_i(x_1, x_2, ..., x_N)$ are the constraints in the system, and $u_i(x_1, x_2, ..., x_N)$ are any smooth enough functions. On the constraints, we get:

$$F(x_1, x_2, ..., x_N) + \sum_{i=1}^{M} u_i(x_1, x_2, ..., x_N) \varphi_i(x_1, x_2, ..., x_N) =$$
$$= F(x_1, x_2, ..., x_N) + \sum_{i=1}^{M} u_i(x_1, x_2, ..., x_N) \cdot 0 =$$
$$= F(x_1, x_2, ..., x_N)$$

Again, we used the fact that any constraint is equal to zero for all solutions of the equations of motion since the solutions must satisfy the constraints.

So, we have:

$$F(x_1, x_2, ..., x_N) + \sum_{i=1}^{M} u_i(x_1, x_2, ..., x_N) \varphi_i(x_1, x_2, ..., x_N) = F(x_1, x_2, ..., x_N) \tag{6.22}$$

This means that also the right side of a differential equation of motion can be modified by any substitution that uses the constraints.

Finally, we look at the left side of the constraint (6.16). Can we modify it using constraints? The calculation here is rather simple, and similar to the previous one. We get:

$$\varphi(x_1, x_2, ..., x_N) + \sum_{i=1}^{M} u_i(x_1, x_2, ..., x_N) \varphi_i(x_1, x_2, ..., x_N) =$$

$$= \varphi(x_1, x_2, ..., x_N) + \sum_{i=1}^{M} u_i(x_1, x_2, ..., x_N) \cdot 0 =$$

$$= \varphi(x_1, x_2, ..., x_N)$$

Again, we used the fact that any constraint was equal to zero for all solutions of the equations of motion, since the solutions must satisfy the constraints.

So, we have:

$$\varphi(x_1, x_2, ..., x_N) + \sum_{i=1}^{M} u_i(x_1, x_2, ..., x_N) \varphi_i(x_1, x_2, ..., x_N) = \varphi(x_1, x_2, ..., x_N) \tag{6.23}$$

To conclude, we can start with any equation of motion, either the differential one or a constraint, and when we apply substitution using a constraint, we will get another, valid equation of motion. It is important that we can use constraints to simplify differential equations of motion or to simplify constraints.

However, it is also important to realize that the equations of motion, which we obtain by substitution using constraints will in general not identically replace the original equations, the one that we started with.

Let us illustrate this with an example. Say we have a system described by the variables (q_1, q_2), subject to three equations of motion, two differential and one constraint. Let the equations of motion be:

$$\begin{aligned} \dot{q}_1 &= 0, \\ \dot{q}_2 &= q_1 - q_2, \\ q_1 - q_2 &= 0. \end{aligned} \tag{6.24}$$

What can we do with this system? First, we can simplify the right side of the second equation, getting:

$$\dot{q}_1 = 0,$$
$$\dot{q}_2 = 0, \tag{6.25}$$
$$q_1 - q_2 = 0.$$

Changing from (6.24) to (6.25) may be beneficial; it simplifies the right side of a differential equations of motion. This can always be done, and the system we obtain is equivalent to the system with which we started.

Now, let us try to use constraints to simplify the left side of the second equation in (6.25). The possible steps are:

$$\frac{d}{dt}(\dot{q}_1 - 0) = 0,$$
$$\dot{q}_2 = 0, \tag{6.26}$$
$$q_1 - q_2 = 0.$$

Then:

$$\frac{d}{dt}(\dot{q}_1 - (q_1 - q_2)) = 0,$$
$$\dot{q}_2 = 0, \tag{6.27}$$
$$q_1 - q_2 = 0.$$

Then:

$$\frac{d}{dt}(q_2) = 0,$$
$$\dot{q}_2 = 0, \tag{6.28}$$
$$q_1 - q_2 = 0.$$

Then:

$$\dot{q}_2 = 0,$$
$$\ddot{q}_2 = 0, \qquad (6.29)$$
$$q_1 - q_2 = 0.$$

Finally, we drop one of the two identical equations of motion (6.29), getting:

$$\dot{q}_2 = 0,$$
$$q_1 - q_2 = 0. \qquad (6.30)$$

Notice that the system (6.30) is still equivalent to (6.24), since the information about the derivative of $\dot{q}_1$ can be obtained back from the fact that q_1 is equal to q_2. We can use (6.30) instead of (6.25) and still retain all the information from (6.25).

So, the equations (6.30) are not identical to the equations (6.24) with which we started. Still, they are equivalent to them.

Even more interestingly, we can rewrite (6.30) as:

$$\dot{q}_2 = 0,$$
$$q_1 = q_2. \qquad (6.31)$$

Then we can forget about q_1 altogether, and just write:

$$\dot{q}_2 = 0, \qquad (6.32)$$

and claim that (6.32) is the only equation describing the system, and q_2 is the only variable describing the system. This claim is validated by the fact that q_1 is completely expressed by q_2, and the time derivative of q_1 is obtained by taking the time derivative of q_1 expressed by q_2. In this sense, q_1 is completely redundant and can be dropped from the description of the system.

This situation is somewhat analogous to having a 2-dimensional plane described by cartesian coordinates and the polar (angle and radius) coordinates simultaneously. So, we are using the variables (x,y,r,ϕ). Without going into specifics, say we have some differential equations for (x,y), and formulas expressing (r,ϕ) by (x,y). These formulas could be interpreted as constraints. Say then that the differential equations for (r,ϕ) would be obtained by taking time derivatives of these constraints.

In this situation it would be completely natural to say that the variables (r,ϕ) are redundant and to drop them from considerations completely. This is analogous to what we did dropping q_1 in (6.31) and switching to (6.32).

Notice that a priori all possibly existing coordinate systems could always be included in our original equations. However, we obviously never use them all (since there exists infinitely many of them), and we usually just choose one system of coordinates.

To conclude, the systems of equations (6.25), (6.30), and (6.32) are equivalent in our approach. Which one we choose is arbitrary, and we pick the one we consider to be the most convenient for the application in front of us.

Finally, maybe an obvious but important warning; when using constraints to simplify constraints, we need to ensure that we get a system equivalent to the original one. If we pay no attention, it is possible to skip an important constraint. For example, say we get a constraint from formula (6.30), namely:

$$q_1 - q_2 = 0. \tag{6.33}$$

We can solve it for q_2, getting:

$$q_2 = q_1. \tag{6.34}$$

Then we can substitute back to (6.33), getting:

$$q_1 - (q_1) = 0, \qquad (6.35)$$

which simplifies to:

$$0 = 0. \qquad (6.36)$$

Obviously, the trivial expression (6.36) cannot replace the original constraint (6.33). Creating trivial expressions by manipulating constraints is always possible, and we need always to make sure when simplifying constraints using other constraints, that we end up with a system that is equivalent to the original one, in the sense that it gives the same restrictions on the allowed points.

CHAPTER VII

POISSON BRACKETS FOR CONSTRAINED SYSTEMS. EFFECTS OF CONSTRAINTS ON THE HAMILTON'S EQUATIONS EXPRESSED BY POISSON BRACKETS

1. Definition of the Poisson Brackets for Systems with Constraints

We can look back at Chapter III, formula (3.1), and Chapter V, formula (5.3), to recall that the Legendre Transformations are identical in constrained and non-constrained cases. We can also see there that in both cases we use the same variables $(q_1, q_2, ..., q_N, p_1, p_2, ..., p_N)$ as position-momenta variables.

Similarly, the definition of Poisson Brackets for a constrained system is going to be identical to the definition of Poisson Brackets for the constrained one. Namely, for an ordered pair of functions $F = F(q_1, q_2, ..., q_N, p_1, p_2, ..., p_N)$ and $G = G(q_1, q_2, ..., q_N, p_1, p_2, ..., p_N)$, we define a new function of $(q_1, q_2, ..., q_N, p_1, p_2, ..., p_N)$ called the Poisson Bracket, denoted by $\{F, G\}$ as:

$$\{F, G\} = \sum_{i=1}^{N} \left(\frac{\partial F}{\partial q_i} \cdot \frac{\partial G}{\partial p_i} - \frac{\partial F}{\partial p_i} \cdot \frac{\partial G}{\partial q_i} \right). \tag{7.1}$$

2. Calculations of Basic Properties of Poisson Brackets for Systems with Constraints

Since the definitions of Poisson Brackets (7.1) and (3.19) are identical for constrained and non-constrained case, then the basic properties of the Poisson Brackets are also identical. They can be found in almost any basic text on Classical Mechanics. We will include it here for completeness.

Assume $E, F,$ and G are some functions of $(q_1, q_2, ..., q_N, p_1, p_2, ..., p_N)$, and a and b are constants. We can perform the following calculations:

a)
$$\{F, G\} = \sum_{i=1}^{N} \left(\frac{\partial F}{\partial q_i} \cdot \frac{\partial G}{\partial p_i} - \frac{\partial F}{\partial p_i} \cdot \frac{\partial G}{\partial q_i} \right) = \sum_{i=1}^{N} \left(-\frac{\partial F}{\partial p_i} \cdot \frac{\partial G}{\partial q_i} + \frac{\partial F}{\partial q_i} \cdot \frac{\partial G}{\partial p_i} \right) =$$

(continued on the next page)

$$= \sum_{i=1}^{N}\left(-\frac{\partial G}{\partial q_i}\cdot\frac{\partial F}{\partial p_i}+\frac{\partial G}{\partial p_i}\cdot\frac{\partial F}{\partial q_i}\right)=\sum_{i=1}^{N}(-1)\left(\frac{\partial G}{\partial q_i}\cdot\frac{\partial F}{\partial p_i}-\frac{\partial G}{\partial p_i}\cdot\frac{\partial F}{\partial q_i}\right)=$$

$$=-\sum_{i=1}^{N}\left(\frac{\partial G}{\partial q_i}\cdot\frac{\partial F}{\partial p_i}-\frac{\partial G}{\partial p_i}\cdot\frac{\partial F}{\partial q_i}\right)=-\{G,F\}$$

b)

$$\{aF,G\}=\sum_{i=1}^{N}\left(\frac{\partial(aF)}{\partial q_i}\cdot\frac{\partial G}{\partial p_i}-\frac{\partial(aF)}{\partial p_i}\cdot\frac{\partial G}{\partial q_i}\right)=\sum_{i=1}^{N}\left(a\cdot\frac{\partial F}{\partial q_i}\cdot\frac{\partial G}{\partial p_i}-a\cdot\frac{\partial F}{\partial p_i}\cdot\frac{\partial G}{\partial q_i}\right)=$$

$$=\sum_{i=1}^{N}a\left(\frac{\partial F}{\partial q_i}\cdot\frac{\partial G}{\partial p_i}-\frac{\partial F}{\partial p_i}\cdot\frac{\partial G}{\partial q_i}\right)=a\sum_{i=1}^{N}\left(\frac{\partial F}{\partial q_i}\cdot\frac{\partial G}{\partial p_i}-\frac{\partial F}{\partial p_i}\cdot\frac{\partial G}{\partial q_i}\right)=$$

$$=a\{F,G\}$$

c) Using the results from a) and b), we get:

$$\{F,aG\}=-\{aG,F\}=-a\{G,F\}=a(-\{G,F\})=$$
$$=a(\{F,G\})=a\{F,G\}$$

d)

$$\{F+G,E\}=\sum_{i=1}^{N}\left(\frac{\partial(F+G)}{\partial q_i}\cdot\frac{\partial E}{\partial p_i}-\frac{\partial(F+G)}{\partial p_i}\cdot\frac{\partial E}{\partial q_i}\right)=$$

$$=\sum_{i=1}^{N}\left(\left(\frac{\partial F}{\partial q_i}+\frac{\partial G}{\partial q_i}\right)\cdot\frac{\partial E}{\partial p_i}-\left(\frac{\partial F}{\partial p_i}+\frac{\partial G}{\partial p_i}\right)\cdot\frac{\partial E}{\partial q_i}\right)=$$

$$=\sum_{i=1}^{N}\left(\frac{\partial F}{\partial q_i}\cdot\frac{\partial E}{\partial p_i}+\frac{\partial G}{\partial q_i}\cdot\frac{\partial E}{\partial p_i}-\frac{\partial F}{\partial p_i}\cdot\frac{\partial E}{\partial q_i}-\frac{\partial G}{\partial p_i}\cdot\frac{\partial E}{\partial q_i}\right)=$$

$$=\sum_{i=1}^{N}\left(\frac{\partial F}{\partial q_i}\cdot\frac{\partial E}{\partial p_i}-\frac{\partial F}{\partial p_i}\cdot\frac{\partial E}{\partial q_i}+\frac{\partial G}{\partial q_i}\cdot\frac{\partial E}{\partial p_i}-\frac{\partial G}{\partial p_i}\cdot\frac{\partial E}{\partial q_i}\right)=$$

$$=\sum_{i=1}^{N}\left(\frac{\partial F}{\partial q_i}\cdot\frac{\partial E}{\partial p_i}-\frac{\partial F}{\partial p_i}\cdot\frac{\partial E}{\partial q_i}\right)+\sum_{i=1}^{N}\left(\frac{\partial G}{\partial q_i}\cdot\frac{\partial E}{\partial p_i}-\frac{\partial G}{\partial p_i}\cdot\frac{\partial E}{\partial q_i}\right)=$$

$$=\{F,E\}+\{G,E\}$$

e) Using the results from a) and d) we get:

$$\{F, G+E\} = -\{G+E, F\} = -(\{G+E, F\}) =$$
$$= -(\{G,F\} + \{E,F\}) = -(-\{F,G\} - \{F,E\}) =$$
$$= \{F,G\} + \{F,E\}$$

f) Using the results from b) and d) we get:

$$\{aF + bG, E\} = \{aF, E\} + \{bG, E\} = a\{F, E\} + b\{G, E\}$$

h) Using the results from c) and e) we get:

$$\{F, aG + bE\} = \{F, aG\} + \{F, bE\} = a\{F, G\} + b\{F, E\}$$

i)

$$\{FG, E\} = \sum_{i=1}^{N} \left(\frac{\partial (FG)}{\partial q_i} \cdot \frac{\partial E}{\partial p_i} - \frac{\partial (FG)}{\partial p_i} \cdot \frac{\partial E}{\partial q_i} \right) =$$

$$= \sum_{i=1}^{N} \left(\left(\frac{\partial F}{\partial q_i} \cdot G + F \cdot \frac{\partial G}{\partial q_i} \right) \cdot \frac{\partial E}{\partial p_i} - \left(\frac{\partial F}{\partial p_i} \cdot G + F \cdot \frac{\partial G}{\partial p_i} \right) \cdot \frac{\partial E}{\partial q_i} \right) =$$

$$= \sum_{i=1}^{N} \left(\frac{\partial F}{\partial q_i} \cdot G \cdot \frac{\partial E}{\partial p_i} + F \cdot \frac{\partial G}{\partial q_i} \cdot \frac{\partial E}{\partial p_i} - \frac{\partial F}{\partial p_i} \cdot G \cdot \frac{\partial E}{\partial q_i} - F \cdot \frac{\partial G}{\partial p_i} \cdot \frac{\partial E}{\partial q_i} \right) =$$

$$= \sum_{i=1}^{N} \left(F \cdot \frac{\partial G}{\partial q_i} \cdot \frac{\partial E}{\partial p_i} - F \cdot \frac{\partial G}{\partial p_i} \cdot \frac{\partial E}{\partial q_i} + \frac{\partial F}{\partial q_i} \cdot G \cdot \frac{\partial E}{\partial p_i} - \frac{\partial F}{\partial p_i} \cdot G \cdot \frac{\partial E}{\partial q_i} \right) =$$

$$= \sum_{i=1}^{N} F \left(\frac{\partial G}{\partial q_i} \cdot \frac{\partial E}{\partial p_i} - \frac{\partial G}{\partial p_i} \cdot \frac{\partial E}{\partial q_i} \right) + \sum_{i=1}^{N} G \left(\frac{\partial F}{\partial q_i} \cdot \frac{\partial E}{\partial p_i} - \frac{\partial F}{\partial p_i} \cdot \frac{\partial E}{\partial q_i} \right) =$$

$$= F \sum_{i=1}^{N} \left(\frac{\partial G}{\partial q_i} \cdot \frac{\partial E}{\partial p_i} - \frac{\partial G}{\partial p_i} \cdot \frac{\partial E}{\partial q_i} \right) + G \sum_{i=1}^{N} \left(\frac{\partial F}{\partial q_i} \cdot \frac{\partial E}{\partial p_i} - \frac{\partial F}{\partial p_i} \cdot \frac{\partial E}{\partial q_i} \right) =$$

$$= F\{G, E\} + G\{F, E\}$$

j) Using the results from a) and i) we get:

$$\{F, GE\} = -\{GE, F\} = -\left(G\{E, F\} + E\{G, F\}\right) =$$
$$= -\left(G\left(-\{F, E\}\right) + E\left(-\{F, G\}\right)\right) = -\left(-G\{F, E\} - E\{F, G\}\right) =$$
$$= G\{F, E\} + E\{F, G\}$$

h) Jacobi Identity

$$\{\{F, G\}, E\} + \{\{G, E\}, F\} + \{\{E, F\}, G\} =$$
$$= \left\{\sum_{i=1}^{N}\left(\frac{\partial F}{\partial q_i} \cdot \frac{\partial G}{\partial p_i} - \frac{\partial F}{\partial p_i} \cdot \frac{\partial G}{\partial q_i}\right), E\right\} + \left\{\sum_{i=1}^{N}\left(\frac{\partial G}{\partial q_i} \cdot \frac{\partial E}{\partial p_i} - \frac{\partial G}{\partial p_i} \cdot \frac{\partial E}{\partial q_i}\right), F\right\} + \left\{\sum_{i=1}^{N}\left(\frac{\partial E}{\partial q_i} \cdot \frac{\partial F}{\partial p_i} - \frac{\partial E}{\partial p_i} \cdot \frac{\partial F}{\partial q_i}\right), G\right\} =$$
$$= \sum_{i=1}^{N}\left\{\frac{\partial F}{\partial q_i} \cdot \frac{\partial G}{\partial p_i} - \frac{\partial F}{\partial p_i} \cdot \frac{\partial G}{\partial q_i}, E\right\} + \sum_{n=1}^{N}\left\{\frac{\partial G}{\partial q_i} \cdot \frac{\partial E}{\partial p_i} - \frac{\partial G}{\partial p_i} \cdot \frac{\partial E}{\partial q_i}, F\right\} + \sum_{n=1}^{N}\left\{\frac{\partial E}{\partial q_i} \cdot \frac{\partial F}{\partial p_i} - \frac{\partial E}{\partial p_i} \cdot \frac{\partial F}{\partial q_i}, G\right\} =$$
$$= \sum_{i=1}^{N}\sum_{j=1}^{N}\left(\frac{\partial\left(\frac{\partial F}{\partial q_i} \cdot \frac{\partial G}{\partial p_i} - \frac{\partial F}{\partial p_i} \cdot \frac{\partial G}{\partial q_i}\right)}{\partial q_j} \cdot \frac{\partial E}{\partial p_j} - \frac{\partial\left(\frac{\partial F}{\partial q_i} \cdot \frac{\partial G}{\partial p_i} - \frac{\partial F}{\partial p_i} \cdot \frac{\partial G}{\partial q_i}\right)}{\partial p_j} \cdot \frac{\partial E}{\partial q_j}\right) +$$
$$+ \sum_{i=1}^{N}\sum_{j=1}^{N}\left(\frac{\partial\left(\frac{\partial G}{\partial q_i} \cdot \frac{\partial E}{\partial p_i} - \frac{\partial G}{\partial p_i} \cdot \frac{\partial E}{\partial q_i}\right)}{\partial q_j} \cdot \frac{\partial F}{\partial p_j} - \frac{\partial\left(\frac{\partial G}{\partial q_i} \cdot \frac{\partial E}{\partial p_i} - \frac{\partial G}{\partial p_i} \cdot \frac{\partial E}{\partial q_i}\right)}{\partial p_j} \cdot \frac{\partial F}{\partial q_j}\right) +$$
$$+ \sum_{i=1}^{N}\sum_{j=1}^{N}\left(\frac{\partial\left(\frac{\partial E}{\partial q_i} \cdot \frac{\partial F}{\partial p_i} - \frac{\partial E}{\partial p_i} \cdot \frac{\partial F}{\partial q_i}\right)}{\partial q_j} \cdot \frac{\partial G}{\partial p_j} - \frac{\partial\left(\frac{\partial E}{\partial q_i} \cdot \frac{\partial F}{\partial p_i} - \frac{\partial E}{\partial p_i} \cdot \frac{\partial F}{\partial q_i}\right)}{\partial p_j} \cdot \frac{\partial G}{\partial q_j}\right) =$$
$$= \sum_{i=1}^{N}\sum_{j=1}^{N}\left(\left(\frac{\partial^2 F}{\partial q_j \partial q_i} \cdot \frac{\partial G}{\partial p_i} + \frac{\partial F}{\partial q_i} \cdot \frac{\partial^2 G}{\partial q_j \partial p_i} - \frac{\partial^2 F}{\partial q_j \partial p_i} \cdot \frac{\partial G}{\partial q_i} - \frac{\partial F}{\partial p_i} \cdot \frac{\partial^2 G}{\partial q_j \partial q_i}\right) \cdot \frac{\partial E}{\partial p_j}\right) +$$
$$- \sum_{i=1}^{N}\sum_{j=1}^{N}\left(\left(\frac{\partial^2 F}{\partial p_j \partial q_i} \cdot \frac{\partial G}{\partial p_i} + \frac{\partial F}{\partial q_i} \cdot \frac{\partial^2 G}{\partial p_j \partial p_i} - \frac{\partial^2 F}{\partial p_j \partial p_i} \cdot \frac{\partial G}{\partial q_i} - \frac{\partial F}{\partial p_i} \cdot \frac{\partial^2 G}{\partial p_j \partial q_i}\right) \cdot \frac{\partial E}{\partial q_j}\right) +$$

(continued on the next page)

$$+\sum_{i=1}^{N}\sum_{j=1}^{N}\left(\left(\frac{\partial^2 G}{\partial q_j \partial q_i}\cdot\frac{\partial E}{\partial p_i}+\frac{\partial G}{\partial q_i}\cdot\frac{\partial^2 E}{\partial q_j \partial p_i}-\frac{\partial^2 G}{\partial q_j \partial p_i}\cdot\frac{\partial E}{\partial q_i}-\frac{\partial G}{\partial p_i}\cdot\frac{\partial^2 E}{\partial q_j \partial q_i}\right)\cdot\frac{\partial F}{\partial p_j}\right)+$$

$$-\sum_{i=1}^{N}\sum_{j=1}^{N}\left(\left(\frac{\partial^2 G}{\partial p_j \partial q_i}\cdot\frac{\partial E}{\partial p_i}+\frac{\partial G}{\partial q_i}\cdot\frac{\partial^2 E}{\partial p_j \partial p_i}-\frac{\partial^2 G}{\partial p_j \partial p_i}\cdot\frac{\partial E}{\partial q_i}-\frac{\partial G}{\partial p_i}\cdot\frac{\partial^2 E}{\partial p_j \partial q_i}\right)\cdot\frac{\partial F}{\partial q_j}\right)+$$

$$+\sum_{i=1}^{N}\sum_{j=1}^{N}\left(\left(\frac{\partial^2 E}{\partial q_j \partial q_i}\cdot\frac{\partial F}{\partial p_i}+\frac{\partial E}{\partial q_i}\cdot\frac{\partial^2 F}{\partial q_j \partial p_i}-\frac{\partial^2 E}{\partial q_j \partial p_i}\cdot\frac{\partial F}{\partial q_i}-\frac{\partial E}{\partial p_i}\cdot\frac{\partial^2 F}{\partial q_j \partial q_i}\right)\cdot\frac{\partial G}{\partial p_j}\right)+$$

$$-\sum_{i=1}^{N}\sum_{j=1}^{N}\left(\left(\frac{\partial^2 E}{\partial p_j \partial q_i}\cdot\frac{\partial F}{\partial p_i}+\frac{\partial E}{\partial q_i}\cdot\frac{\partial^2 F}{\partial p_j \partial p_i}-\frac{\partial^2 E}{\partial p_j \partial p_i}\cdot\frac{\partial F}{\partial q_i}-\frac{\partial E}{\partial p_i}\cdot\frac{\partial^2 F}{\partial p_j \partial q_i}\right)\cdot\frac{\partial G}{\partial q_j}\right)=$$

using the distribution:

$$=\sum_{i=1}^{N}\sum_{j=1}^{N}\left(\frac{\partial^2 F}{\partial q_j \partial q_i}\cdot\frac{\partial G}{\partial p_i}\cdot\frac{\partial E}{\partial p_j}+\frac{\partial F}{\partial q_i}\cdot\frac{\partial^2 G}{\partial q_j \partial p_i}\cdot\frac{\partial E}{\partial p_j}-\frac{\partial^2 F}{\partial q_j \partial p_i}\cdot\frac{\partial G}{\partial q_i}\cdot\frac{\partial E}{\partial p_j}-\frac{\partial F}{\partial p_i}\cdot\frac{\partial^2 G}{\partial q_j \partial q_i}\cdot\frac{\partial E}{\partial p_j}\right)+$$

$$-\sum_{i=1}^{N}\sum_{j=1}^{N}\left(\frac{\partial^2 F}{\partial p_j \partial q_i}\cdot\frac{\partial G}{\partial p_i}\cdot\frac{\partial E}{\partial q_j}+\frac{\partial F}{\partial q_i}\cdot\frac{\partial^2 G}{\partial p_j \partial p_i}\cdot\frac{\partial E}{\partial q_j}-\frac{\partial^2 F}{\partial p_j \partial p_i}\cdot\frac{\partial G}{\partial q_i}\cdot\frac{\partial E}{\partial q_j}-\frac{\partial F}{\partial p_i}\cdot\frac{\partial^2 G}{\partial p_j \partial q_i}\cdot\frac{\partial E}{\partial q_j}\right)+$$

$$+\sum_{i=1}^{N}\sum_{j=1}^{N}\left(\frac{\partial^2 G}{\partial q_j \partial q_i}\cdot\frac{\partial E}{\partial p_i}\cdot\frac{\partial F}{\partial p_j}+\frac{\partial G}{\partial q_i}\cdot\frac{\partial^2 E}{\partial q_j \partial p_i}\cdot\frac{\partial F}{\partial p_j}-\frac{\partial^2 G}{\partial q_j \partial p_i}\cdot\frac{\partial E}{\partial q_i}\cdot\frac{\partial F}{\partial p_j}-\frac{\partial G}{\partial p_i}\cdot\frac{\partial^2 E}{\partial q_j \partial q_i}\cdot\frac{\partial F}{\partial p_j}\right)+$$

$$-\sum_{i=1}^{N}\sum_{j=1}^{N}\left(\frac{\partial^2 G}{\partial p_j \partial q_i}\cdot\frac{\partial E}{\partial p_i}\cdot\frac{\partial F}{\partial q_j}+\frac{\partial G}{\partial q_i}\cdot\frac{\partial^2 E}{\partial p_j \partial p_i}\cdot\frac{\partial F}{\partial q_j}-\frac{\partial^2 G}{\partial p_j \partial p_i}\cdot\frac{\partial E}{\partial q_i}\cdot\frac{\partial F}{\partial q_j}-\frac{\partial G}{\partial p_i}\cdot\frac{\partial^2 E}{\partial p_j \partial q_i}\cdot\frac{\partial F}{\partial q_j}\right)+$$

$$+\sum_{i=1}^{N}\sum_{j=1}^{N}\left(\frac{\partial^2 E}{\partial q_j \partial q_i}\cdot\frac{\partial F}{\partial p_i}\cdot\frac{\partial G}{\partial p_j}+\frac{\partial E}{\partial q_i}\cdot\frac{\partial^2 F}{\partial q_j \partial p_i}\cdot\frac{\partial G}{\partial p_j}-\frac{\partial^2 E}{\partial q_j \partial p_i}\cdot\frac{\partial F}{\partial q_i}\cdot\frac{\partial G}{\partial p_j}-\frac{\partial E}{\partial p_i}\cdot\frac{\partial^2 F}{\partial q_j \partial q_i}\cdot\frac{\partial G}{\partial p_j}\right)+$$

$$-\sum_{i=1}^{N}\sum_{j=1}^{N}\left(\frac{\partial^2 E}{\partial p_j \partial q_i}\cdot\frac{\partial F}{\partial p_i}\cdot\frac{\partial G}{\partial q_j}+\frac{\partial E}{\partial q_i}\cdot\frac{\partial^2 F}{\partial p_j \partial p_i}\cdot\frac{\partial G}{\partial q_j}-\frac{\partial^2 E}{\partial p_j \partial p_i}\cdot\frac{\partial F}{\partial q_i}\cdot\frac{\partial G}{\partial q_j}-\frac{\partial E}{\partial p_i}\cdot\frac{\partial^2 F}{\partial p_j \partial q_i}\cdot\frac{\partial G}{\partial q_j}\right)=$$

distributing the minuses in front of summations:

$$=\sum_{i=1}^{N}\sum_{j=1}^{N}\left(\frac{\partial^2 F}{\partial q_j \partial q_i}\cdot\frac{\partial G}{\partial p_i}\cdot\frac{\partial E}{\partial p_j}+\frac{\partial F}{\partial q_i}\cdot\frac{\partial^2 G}{\partial q_j \partial p_i}\cdot\frac{\partial E}{\partial p_j}-\frac{\partial^2 F}{\partial q_j \partial p_i}\cdot\frac{\partial G}{\partial q_i}\cdot\frac{\partial E}{\partial p_j}-\frac{\partial F}{\partial p_i}\cdot\frac{\partial^2 G}{\partial q_j \partial q_i}\cdot\frac{\partial E}{\partial p_j}\right)+$$

(continued on the next page)

$$+\sum_{i=1}^{N}\sum_{j=1}^{N}\left(-\frac{\partial^2 F}{\partial p_j \partial q_i}\cdot\frac{\partial G}{\partial p_i}\cdot\frac{\partial E}{\partial q_j}-\frac{\partial F}{\partial q_i}\cdot\frac{\partial^2 G}{\partial p_j \partial p_i}\cdot\frac{\partial E}{\partial q_j}+\frac{\partial^2 F}{\partial p_j \partial p_i}\cdot\frac{\partial G}{\partial q_i}\cdot\frac{\partial E}{\partial q_j}+\frac{\partial F}{\partial p_i}\cdot\frac{\partial^2 G}{\partial p_j \partial q_i}\cdot\frac{\partial E}{\partial q_j}\right)+$$

$$+\sum_{i=1}^{N}\sum_{j=1}^{N}\left(\frac{\partial^2 G}{\partial q_j \partial q_i}\cdot\frac{\partial E}{\partial p_i}\cdot\frac{\partial F}{\partial p_j}+\frac{\partial G}{\partial q_i}\cdot\frac{\partial^2 E}{\partial q_j \partial p_i}\cdot\frac{\partial F}{\partial p_j}-\frac{\partial^2 G}{\partial q_j \partial p_i}\cdot\frac{\partial E}{\partial q_i}\cdot\frac{\partial F}{\partial p_j}-\frac{\partial G}{\partial p_i}\cdot\frac{\partial^2 E}{\partial q_j \partial q_i}\cdot\frac{\partial F}{\partial p_j}\right)+$$

$$+\sum_{i=1}^{N}\sum_{j=1}^{N}\left(-\frac{\partial^2 G}{\partial p_j \partial q_i}\cdot\frac{\partial E}{\partial p_i}\cdot\frac{\partial F}{\partial q_j}-\frac{\partial G}{\partial q_i}\cdot\frac{\partial^2 E}{\partial p_j \partial p_i}\cdot\frac{\partial F}{\partial q_j}+\frac{\partial^2 G}{\partial p_j \partial p_i}\cdot\frac{\partial E}{\partial q_i}\cdot\frac{\partial F}{\partial q_j}+\frac{\partial G}{\partial p_i}\cdot\frac{\partial^2 E}{\partial p_j \partial q_i}\cdot\frac{\partial F}{\partial q_j}\right)+$$

$$+\sum_{i=1}^{N}\sum_{j=1}^{N}\left(\frac{\partial^2 E}{\partial q_j \partial q_i}\cdot\frac{\partial F}{\partial p_i}\cdot\frac{\partial G}{\partial p_j}+\frac{\partial E}{\partial q_i}\cdot\frac{\partial^2 F}{\partial q_j \partial p_i}\cdot\frac{\partial G}{\partial p_j}-\frac{\partial^2 E}{\partial q_j \partial p_i}\cdot\frac{\partial F}{\partial q_i}\cdot\frac{\partial G}{\partial p_j}-\frac{\partial E}{\partial p_i}\cdot\frac{\partial^2 F}{\partial q_j \partial q_i}\cdot\frac{\partial G}{\partial p_j}\right)+$$

$$+\sum_{i=1}^{N}\sum_{j=1}^{N}\left(-\frac{\partial^2 E}{\partial p_j \partial q_i}\cdot\frac{\partial F}{\partial p_i}\cdot\frac{\partial G}{\partial q_j}-\frac{\partial E}{\partial q_i}\cdot\frac{\partial^2 F}{\partial p_j \partial p_i}\cdot\frac{\partial G}{\partial q_j}+\frac{\partial^2 E}{\partial p_j \partial p_i}\cdot\frac{\partial F}{\partial q_i}\cdot\frac{\partial G}{\partial q_j}+\frac{\partial E}{\partial p_i}\cdot\frac{\partial^2 F}{\partial p_j \partial q_i}\cdot\frac{\partial G}{\partial q_j}\right)=$$

combining all summations:

$$=\sum_{i=1}^{N}\sum_{j=1}^{N}\left(\frac{\partial^2 F}{\partial q_j \partial q_i}\cdot\frac{\partial G}{\partial p_i}\cdot\frac{\partial E}{\partial p_j}+\frac{\partial F}{\partial q_i}\cdot\frac{\partial^2 G}{\partial q_j \partial p_i}\cdot\frac{\partial E}{\partial p_j}-\frac{\partial^2 F}{\partial q_j \partial p_i}\cdot\frac{\partial G}{\partial q_i}\cdot\frac{\partial E}{\partial p_j}-\frac{\partial F}{\partial p_i}\cdot\frac{\partial^2 G}{\partial q_j \partial q_i}\cdot\frac{\partial E}{\partial p_j}+\right.$$

$$-\frac{\partial^2 F}{\partial p_j \partial q_i}\cdot\frac{\partial G}{\partial p_i}\cdot\frac{\partial E}{\partial q_j}-\frac{\partial F}{\partial q_i}\cdot\frac{\partial^2 G}{\partial p_j \partial p_i}\cdot\frac{\partial E}{\partial q_j}+\frac{\partial^2 F}{\partial p_j \partial p_i}\cdot\frac{\partial G}{\partial q_i}\cdot\frac{\partial E}{\partial q_j}+\frac{\partial F}{\partial p_i}\cdot\frac{\partial^2 G}{\partial p_j \partial q_i}\cdot\frac{\partial E}{\partial q_j}+$$

$$+\frac{\partial^2 G}{\partial q_j \partial q_i}\cdot\frac{\partial E}{\partial p_i}\cdot\frac{\partial F}{\partial p_j}+\frac{\partial G}{\partial q_i}\cdot\frac{\partial^2 E}{\partial q_j \partial p_i}\cdot\frac{\partial F}{\partial p_j}-\frac{\partial^2 G}{\partial q_j \partial p_i}\cdot\frac{\partial E}{\partial q_i}\cdot\frac{\partial F}{\partial p_j}-\frac{\partial G}{\partial p_i}\cdot\frac{\partial^2 E}{\partial q_j \partial q_i}\cdot\frac{\partial F}{\partial p_j}+$$

$$-\frac{\partial^2 G}{\partial p_j \partial q_i}\cdot\frac{\partial E}{\partial p_i}\cdot\frac{\partial F}{\partial q_j}-\frac{\partial G}{\partial q_i}\cdot\frac{\partial^2 E}{\partial p_j \partial p_i}\cdot\frac{\partial F}{\partial q_j}+\frac{\partial^2 G}{\partial p_j \partial p_i}\cdot\frac{\partial E}{\partial q_i}\cdot\frac{\partial F}{\partial q_j}+\frac{\partial G}{\partial p_i}\cdot\frac{\partial^2 E}{\partial p_j \partial q_i}\cdot\frac{\partial F}{\partial q_j}+$$

$$+\frac{\partial^2 E}{\partial q_j \partial q_i}\cdot\frac{\partial F}{\partial p_i}\cdot\frac{\partial G}{\partial p_j}+\frac{\partial E}{\partial q_i}\cdot\frac{\partial^2 F}{\partial q_j \partial p_i}\cdot\frac{\partial G}{\partial p_j}-\frac{\partial^2 E}{\partial q_j \partial p_i}\cdot\frac{\partial F}{\partial q_i}\cdot\frac{\partial G}{\partial p_j}-\frac{\partial E}{\partial p_i}\cdot\frac{\partial^2 F}{\partial q_j \partial q_i}\cdot\frac{\partial G}{\partial p_j}+$$

$$\left.-\frac{\partial^2 E}{\partial p_j \partial q_i}\cdot\frac{\partial F}{\partial p_i}\cdot\frac{\partial G}{\partial q_j}-\frac{\partial E}{\partial q_i}\cdot\frac{\partial^2 F}{\partial p_j \partial p_i}\cdot\frac{\partial G}{\partial q_j}+\frac{\partial^2 E}{\partial p_j \partial p_i}\cdot\frac{\partial F}{\partial q_i}\cdot\frac{\partial G}{\partial q_j}+\frac{\partial E}{\partial p_i}\cdot\frac{\partial^2 F}{\partial p_j \partial q_i}\cdot\frac{\partial G}{\partial q_j}\right)=$$

changing the order, so nearby terms include the second derivative of the same function:

$$=\sum_{i=1}^{N}\sum_{j=1}^{N}\left(\frac{\partial^2 F}{\partial q_j \partial q_i}\cdot\frac{\partial G}{\partial p_i}\cdot\frac{\partial E}{\partial p_j}-\frac{\partial^2 F}{\partial q_j \partial p_i}\cdot\frac{\partial G}{\partial q_i}\cdot\frac{\partial E}{\partial p_j}-\frac{\partial^2 F}{\partial p_j \partial q_i}\cdot\frac{\partial G}{\partial p_i}\cdot\frac{\partial E}{\partial q_j}+\frac{\partial^2 F}{\partial p_j \partial p_i}\cdot\frac{\partial G}{\partial q_i}\cdot\frac{\partial E}{\partial q_j}+\right.$$

(continued on the next page)

$$+\frac{\partial E}{\partial q_i}\cdot\frac{\partial^2 F}{\partial q_j \partial p_i}\cdot\frac{\partial G}{\partial p_j} - \frac{\partial E}{\partial p_i}\cdot\frac{\partial^2 F}{\partial q_j \partial q_i}\cdot\frac{\partial G}{\partial p_j} - \frac{\partial E}{\partial q_i}\cdot\frac{\partial^2 F}{\partial p_j \partial p_i}\cdot\frac{\partial G}{\partial q_j} + \frac{\partial E}{\partial p_i}\cdot\frac{\partial^2 F}{\partial p_j \partial q_i}\cdot\frac{\partial G}{\partial q_j} +$$

$$+\frac{\partial F}{\partial q_i}\cdot\frac{\partial^2 G}{\partial q_j \partial p_i}\cdot\frac{\partial E}{\partial p_j} - \frac{\partial F}{\partial p_i}\cdot\frac{\partial^2 G}{\partial q_j \partial q_i}\cdot\frac{\partial E}{\partial p_j} - \frac{\partial F}{\partial q_i}\cdot\frac{\partial^2 G}{\partial p_j \partial p_i}\cdot\frac{\partial E}{\partial q_j} + \frac{\partial F}{\partial p_i}\cdot\frac{\partial^2 G}{\partial p_j \partial q_i}\cdot\frac{\partial E}{\partial q_j} +$$

$$+\frac{\partial^2 G}{\partial q_j \partial q_i}\cdot\frac{\partial E}{\partial p_i}\cdot\frac{\partial F}{\partial p_j} - \frac{\partial^2 G}{\partial q_j \partial p_i}\cdot\frac{\partial E}{\partial q_i}\cdot\frac{\partial F}{\partial p_j} - \frac{\partial^2 G}{\partial p_j \partial q_i}\cdot\frac{\partial E}{\partial p_i}\cdot\frac{\partial F}{\partial q_j} + \frac{\partial^2 G}{\partial p_j \partial p_i}\cdot\frac{\partial E}{\partial q_i}\cdot\frac{\partial F}{\partial q_j} +$$

$$+\frac{\partial G}{\partial q_i}\cdot\frac{\partial^2 E}{\partial q_j \partial p_i}\cdot\frac{\partial F}{\partial p_j} - \frac{\partial G}{\partial p_i}\cdot\frac{\partial^2 E}{\partial q_j \partial q_i}\cdot\frac{\partial F}{\partial p_j} - \frac{\partial G}{\partial q_i}\cdot\frac{\partial^2 E}{\partial p_j \partial p_i}\cdot\frac{\partial F}{\partial q_j} + \frac{\partial G}{\partial p_i}\cdot\frac{\partial^2 E}{\partial p_j \partial q_i}\cdot\frac{\partial F}{\partial q_j} +$$

$$+\frac{\partial^2 E}{\partial q_j \partial q_i}\cdot\frac{\partial F}{\partial p_i}\cdot\frac{\partial G}{\partial p_j} - \frac{\partial^2 E}{\partial q_j \partial p_i}\cdot\frac{\partial F}{\partial q_i}\cdot\frac{\partial G}{\partial p_j} - \frac{\partial^2 E}{\partial p_j \partial q_i}\cdot\frac{\partial F}{\partial p_i}\cdot\frac{\partial G}{\partial q_j} + \frac{\partial^2 E}{\partial p_j \partial p_i}\cdot\frac{\partial F}{\partial q_i}\cdot\frac{\partial G}{\partial q_j}\Bigg) =$$

placing second derivative at the beginning of each term and writing like terms next to each other:

$$= \sum_{i=1}^{N}\sum_{j=1}^{N}\Bigg(\frac{\partial^2 F}{\partial q_j \partial q_i}\cdot\frac{\partial G}{\partial p_i}\cdot\frac{\partial E}{\partial p_j} - \frac{\partial^2 F}{\partial q_j \partial q_i}\cdot\frac{\partial G}{\partial p_j}\cdot\frac{\partial E}{\partial p_i} - \frac{\partial^2 F}{\partial q_j \partial p_i}\cdot\frac{\partial G}{\partial q_i}\cdot\frac{\partial E}{\partial p_j} + \frac{\partial^2 F}{\partial p_j \partial q_i}\cdot\frac{\partial G}{\partial q_j}\cdot\frac{\partial E}{\partial p_i} +$$

$$+\frac{\partial^2 F}{\partial q_j \partial p_i}\cdot\frac{\partial G}{\partial p_j}\cdot\frac{\partial E}{\partial q_i} - \frac{\partial^2 F}{\partial p_j \partial q_i}\cdot\frac{\partial G}{\partial p_i}\cdot\frac{\partial E}{\partial q_j} - \frac{\partial^2 F}{\partial p_j \partial p_i}\cdot\frac{\partial G}{\partial q_j}\cdot\frac{\partial E}{\partial q_i} + \frac{\partial^2 F}{\partial p_j \partial p_i}\cdot\frac{\partial G}{\partial q_i}\cdot\frac{\partial E}{\partial q_j} +$$

$$+\frac{\partial^2 G}{\partial q_j \partial p_i}\cdot\frac{\partial F}{\partial q_i}\cdot\frac{\partial E}{\partial p_j} - \frac{\partial^2 G}{\partial p_j \partial q_i}\cdot\frac{\partial F}{\partial q_j}\cdot\frac{\partial E}{\partial p_i} - \frac{\partial^2 G}{\partial q_j \partial q_i}\cdot\frac{\partial F}{\partial p_i}\cdot\frac{\partial E}{\partial p_j} + \frac{\partial^2 G}{\partial q_j \partial q_i}\cdot\frac{\partial F}{\partial p_j}\cdot\frac{\partial E}{\partial p_i} +$$

$$-\frac{\partial^2 G}{\partial q_j \partial p_i}\cdot\frac{\partial F}{\partial p_j}\cdot\frac{\partial E}{\partial q_i} + \frac{\partial^2 G}{\partial p_j \partial q_i}\cdot\frac{\partial F}{\partial p_i}\cdot\frac{\partial E}{\partial q_j} + \frac{\partial^2 G}{\partial p_j \partial p_i}\cdot\frac{\partial F}{\partial q_j}\cdot\frac{\partial E}{\partial q_i} - \frac{\partial^2 G}{\partial p_j \partial p_i}\cdot\frac{\partial F}{\partial q_i}\cdot\frac{\partial E}{\partial q_j} +$$

$$+\frac{\partial^2 E}{\partial q_j \partial p_i}\cdot\frac{\partial G}{\partial q_i}\cdot\frac{\partial F}{\partial p_j} - \frac{\partial^2 E}{\partial p_j \partial q_i}\cdot\frac{\partial G}{\partial q_j}\cdot\frac{\partial F}{\partial p_i} - \frac{\partial^2 E}{\partial q_j \partial q_i}\cdot\frac{\partial F}{\partial p_i}\cdot\frac{\partial G}{\partial p_j} + \frac{\partial^2 E}{\partial q_j \partial q_i}\cdot\frac{\partial F}{\partial p_j}\cdot\frac{\partial G}{\partial p_i} +$$

$$-\frac{\partial^2 E}{\partial q_j \partial p_i}\cdot\frac{\partial F}{\partial q_i}\cdot\frac{\partial G}{\partial p_j} + \frac{\partial^2 E}{\partial p_j \partial q_i}\cdot\frac{\partial F}{\partial q_j}\cdot\frac{\partial G}{\partial p_i} + \frac{\partial^2 E}{\partial p_j \partial p_i}\cdot\frac{\partial F}{\partial q_i}\cdot\frac{\partial G}{\partial q_j} - \frac{\partial^2 E}{\partial p_j \partial p_i}\cdot\frac{\partial F}{\partial q_j}\cdot\frac{\partial G}{\partial q_i}\Bigg) = 0.$$

The above expression simplifies to zero, since each consecutive pair in it equals zero. Notice that the indexes i and j can be switched, since they are dummy variables. Also, the order of partial derivatives may be changed, canceling out the terms of each pair.

3. The Poisson Brackets Expressed in Different Sets of Coordinates

The Poisson Brackets were defined in formula (7.1) as:

$$\{F,G\} = \sum_{i=1}^{N} \left(\frac{\partial F}{\partial q_i} \cdot \frac{\partial G}{\partial p_i} - \frac{\partial F}{\partial p_i} \cdot \frac{\partial G}{\partial q_i} \right).$$

In some situations, it may be more useful to use variables different than (q, p). To change to these new variables, it is convenient to rewrite the formula (7.1) using matrices. It is easy to see that the formula (7.1) can be rewritten as matrix multiplication:

$$\{F,G\} = \left(\frac{\partial F}{\partial q_1}, \ldots, \frac{\partial F}{\partial q_N}, \frac{\partial F}{\partial p_1}, \ldots, \frac{\partial F}{\partial p_N} \right) [J] \begin{pmatrix} \frac{\partial G}{\partial q_1} \\ \vdots \\ \frac{\partial G}{\partial q_N} \\ \frac{\partial G}{\partial p_1} \\ \vdots \\ \frac{\partial G}{\partial p_N} \end{pmatrix}, \qquad (7.2)$$

where the matrix $[J]$ is defined as:

$$[J] = \left(\begin{array}{ccc|ccc} 0 & \cdots \cdots & 0 & 1 & 0 & \cdots & 0 \\ \vdots & \ddots & \vdots & 0 & \ddots & & \vdots \\ \vdots & & \ddots & \vdots & & \ddots & 0 \\ 0 & \cdots \cdots & 0 & 0 & \cdots & \cdots & 1 \\ \hline -1 & 0 & \cdots & 0 & 0 & \cdots & \cdots & 0 \\ 0 & \ddots & & \vdots & \vdots & \ddots & & \vdots \\ \vdots & & \ddots & 0 & \vdots & & \ddots & \vdots \\ 0 & \cdots & \cdots & -1 & 0 & \cdots & \cdots & 0 \end{array} \right), \qquad (7.3)$$

Let us now make a change of variables. Say, we call the new variables $(x_1, x_2, ..., x_{2N})$. The change of variables formulas can be written as:

$$\begin{aligned} x_1 &= x_1(q,p), \\ x_2 &= x_2(q,p), \\ &\vdots \\ x_{2N} &= x_{2N}(q,p). \end{aligned} \quad (7.4)$$

Since the (7.4) is the change of variables, it must be invertible. We will denote the inverse by:

$$\begin{aligned} q_1 &= q_1(x_1,...,x_{2N}), \\ &\vdots \\ q_N &= q_N(x_1,...,x_{2N}), \\ p_1 &= p_1(x_1,...,x_{2N}), \\ &\vdots \\ p_N &= p_N(x_1,...,x_{2N}). \end{aligned} \quad (7.5)$$

Then, the chain rule allows us to rewrite (7.5) as:

$$\begin{bmatrix} \dfrac{\partial G}{\partial q} \\ \dfrac{\partial G}{\partial p} \end{bmatrix} = \begin{pmatrix} \dfrac{\partial G}{\partial q_1} \\ \vdots \\ \dfrac{\partial G}{\partial q_N} \\ \dfrac{\partial G}{\partial p_1} \\ \vdots \\ \dfrac{\partial G}{\partial p_N} \end{pmatrix} = \begin{pmatrix} \sum_{k=1}^{2N} \dfrac{\partial x_k}{\partial q_1} \cdot \dfrac{\partial G}{\partial x_k} \\ \sum_{k=1}^{2N} \dfrac{\partial x_k}{\partial q_2} \cdot \dfrac{\partial G}{\partial x_k} \\ \vdots \\ \sum_{k=1}^{2N} \dfrac{\partial x_k}{\partial q_N} \cdot \dfrac{\partial G}{\partial x_k} \\ \sum_{k=1}^{2N} \dfrac{\partial x_k}{\partial p_1} \cdot \dfrac{\partial G}{\partial x_k} \\ \sum_{k=1}^{2N} \dfrac{\partial x_k}{\partial p_2} \cdot \dfrac{\partial G}{\partial x_k} \\ \vdots \\ \sum_{k=1}^{2N} \dfrac{\partial x_k}{\partial p_N} \cdot \dfrac{\partial G}{\partial x_k} \end{pmatrix} \quad (7.6)$$

The matrix on the right side of (7.7) can be written as a multiplication of the matrices:

$$\begin{pmatrix} \sum_{k=1}^{2N} \frac{\partial x_k}{\partial q_1} \cdot \frac{\partial G}{\partial x_k} \\ \sum_{k=1}^{2N} \frac{\partial x_k}{\partial q_2} \cdot \frac{\partial G}{\partial x_k} \\ \vdots \\ \sum_{k=1}^{2N} \frac{\partial x_k}{\partial q_N} \cdot \frac{\partial G}{\partial x_k} \\ \sum_{k=1}^{2N} \frac{\partial x_k}{\partial p_1} \cdot \frac{\partial G}{\partial x_k} \\ \sum_{k=1}^{2N} \frac{\partial x_k}{\partial p_2} \cdot \frac{\partial G}{\partial x_k} \\ \vdots \\ \sum_{k=1}^{2N} \frac{\partial x_k}{\partial p_N} \cdot \frac{\partial G}{\partial x_k} \end{pmatrix} = \begin{pmatrix} \frac{\partial x_1}{\partial q_1} & \frac{\partial x_2}{\partial q_1} & \cdots & \frac{\partial x_{2N}}{\partial q_1} \\ \frac{\partial x_1}{\partial q_2} & \frac{\partial x_2}{\partial q_2} & \cdots & \frac{\partial x_{2N}}{\partial q_2} \\ \vdots & \vdots & & \vdots \\ \frac{\partial x_1}{\partial q_N} & \frac{\partial x_2}{\partial q_N} & \cdots & \frac{\partial x_{2N}}{\partial q_N} \\ \frac{\partial x_1}{\partial p_1} & \frac{\partial x_2}{\partial p_1} & \cdots & \frac{\partial x_{2N}}{\partial p_1} \\ \frac{\partial x_1}{\partial p_2} & \frac{\partial x_2}{\partial p_2} & \cdots & \frac{\partial x_{2N}}{\partial p_2} \\ \vdots & \vdots & & \vdots \\ \frac{\partial x_1}{\partial p_N} & \frac{\partial x_2}{\partial p_N} & \cdots & \frac{\partial x_{2N}}{\partial p_N} \end{pmatrix} \begin{pmatrix} \frac{\partial G}{\partial x_1} \\ \frac{\partial G}{\partial x_2} \\ \vdots \\ \frac{\partial G}{\partial x_{2N}} \end{pmatrix}.$$

(7.7)

So, if we define the matrix:

$$\begin{bmatrix} \frac{\partial x}{\partial q} \\ \frac{\partial x}{\partial p} \end{bmatrix} = \begin{pmatrix} \frac{\partial x_1}{\partial q_1} & \frac{\partial x_2}{\partial q_1} & \cdots & \frac{\partial x_{2N}}{\partial q_1} \\ \frac{\partial x_1}{\partial q_2} & \frac{\partial x_2}{\partial q_2} & \cdots & \frac{\partial x_{2N}}{\partial q_2} \\ \vdots & \vdots & & \vdots \\ \frac{\partial x_1}{\partial q_N} & \frac{\partial x_2}{\partial q_N} & \cdots & \frac{\partial x_{2N}}{\partial q_N} \\ \frac{\partial x_1}{\partial p_1} & \frac{\partial x_2}{\partial p_1} & \cdots & \frac{\partial x_{2N}}{\partial p_1} \\ \frac{\partial x_1}{\partial p_2} & \frac{\partial x_2}{\partial p_2} & \cdots & \frac{\partial x_{2N}}{\partial p_2} \\ \vdots & \vdots & & \vdots \\ \frac{\partial x_1}{\partial p_N} & \frac{\partial x_2}{\partial p_N} & \cdots & \frac{\partial x_{2N}}{\partial p_N} \end{pmatrix},$$

(7.8)

we get:

$$\begin{pmatrix} \dfrac{\partial G}{\partial q_1} \\ \vdots \\ \dfrac{\partial G}{\partial q_N} \\ \dfrac{\partial G}{\partial p_1} \\ \vdots \\ \dfrac{\partial G}{\partial p_N} \end{pmatrix} = \begin{bmatrix} \dfrac{\partial x}{\partial q} \\ \dfrac{\partial x}{\partial p} \end{bmatrix} \begin{pmatrix} \dfrac{\partial G}{\partial x_1} \\ \dfrac{\partial G}{\partial x_2} \\ \vdots \\ \dfrac{\partial G}{\partial x_{2N}} \end{pmatrix}. \qquad (7.9)$$

Similarly, we have

$$\left[\dfrac{\partial F}{\partial q}, \dfrac{\partial F}{\partial p}\right] = \left(\dfrac{\partial F}{\partial q_1}, ..., \dfrac{\partial F}{\partial q_N}, \dfrac{\partial F}{\partial p_1}, ..., \dfrac{\partial F}{\partial p_N}\right) =$$
$$= \left(\sum_{k=1}^{2N} \dfrac{\partial x_k}{\partial q_1} \cdot \dfrac{\partial F}{\partial x_k}, \sum_{k=1}^{2N} \dfrac{\partial x_k}{\partial q_2} \cdot \dfrac{\partial F}{\partial x_k}, ..., \sum_{k=1}^{2N} \dfrac{\partial x_k}{\partial q_N} \cdot \dfrac{\partial F}{\partial x_k}, \sum_{k=1}^{2N} \dfrac{\partial x_k}{\partial p_1} \cdot \dfrac{\partial F}{\partial x_k}, \sum_{k=1}^{2N} \dfrac{\partial x_k}{\partial p_2} \cdot \dfrac{\partial F}{\partial x_k}, ..., \sum_{k=1}^{2N} \dfrac{\partial x_k}{\partial p_N} \cdot \dfrac{\partial F}{\partial x_k}\right). \qquad (7.10)$$

The last line in (7.10) can be rewritten as matrix multiplication:

$$\left(\sum_{k=1}^{2N} \dfrac{\partial x_k}{\partial q_1} \cdot \dfrac{\partial F}{\partial x_k}, \sum_{k=1}^{2N} \dfrac{\partial x_k}{\partial q_2} \cdot \dfrac{\partial F}{\partial x_k}, ..., \sum_{k=1}^{2N} \dfrac{\partial x_k}{\partial q_N} \cdot \dfrac{\partial F}{\partial x_k}, \sum_{k=1}^{2N} \dfrac{\partial x_k}{\partial p_1} \cdot \dfrac{\partial F}{\partial x_k}, \sum_{k=1}^{2N} \dfrac{\partial x_k}{\partial p_2} \cdot \dfrac{\partial F}{\partial x_k}, ..., \sum_{k=1}^{2N} \dfrac{\partial x_k}{\partial p_N} \cdot \dfrac{\partial F}{\partial x_k}\right) =$$
$$= \left(\dfrac{\partial F}{\partial x_1}, \dfrac{\partial F}{\partial x_2}, ..., \dfrac{\partial F}{\partial x_{2N}}\right) \begin{pmatrix} \dfrac{\partial x_1}{\partial q_1} & \dfrac{\partial x_1}{\partial q_2} & \cdots & \dfrac{\partial x_1}{\partial q_N} & \dfrac{\partial x_1}{\partial p_1} & \dfrac{\partial x_1}{\partial p_2} & \cdots & \dfrac{\partial x_1}{\partial p_N} \\ \dfrac{\partial x_2}{\partial q_1} & \dfrac{\partial x_2}{\partial q_2} & \cdots & \dfrac{\partial x_2}{\partial q_N} & \dfrac{\partial x_2}{\partial p_1} & \dfrac{\partial x_2}{\partial p_2} & \cdots & \dfrac{\partial x_2}{\partial p_N} \\ \vdots & \vdots & & \vdots & \vdots & \vdots & & \vdots \\ \dfrac{\partial x_{2N}}{\partial q_1} & \dfrac{\partial x_{2N}}{\partial q_2} & \cdots & \dfrac{\partial x_{2N}}{\partial q_N} & \dfrac{\partial x_{2N}}{\partial p_1} & \dfrac{\partial x_{2N}}{\partial p_2} & \cdots & \dfrac{\partial x_{2N}}{\partial p_N} \end{pmatrix}. \qquad (7.11)$$

Defining the matrix:

$$\left[\frac{\partial x}{\partial q} \quad \frac{\partial x}{\partial p}\right] = \begin{pmatrix} \frac{\partial x_1}{\partial q_1} & \frac{\partial x_1}{\partial q_2} & \cdots & \frac{\partial x_1}{\partial q_N} & \frac{\partial x_1}{\partial p_1} & \frac{\partial x_1}{\partial p_2} & \cdots & \frac{\partial x_1}{\partial p_N} \\ \frac{\partial x_2}{\partial q_1} & \frac{\partial x_2}{\partial q_2} & \cdots & \frac{\partial x_2}{\partial q_N} & \frac{\partial x_2}{\partial p_1} & \frac{\partial x_2}{\partial p_2} & \cdots & \frac{\partial x_2}{\partial p_N} \\ \vdots & \vdots & & \vdots & \vdots & \vdots & & \vdots \\ \frac{\partial x_{2N}}{\partial q_1} & \frac{\partial x_{2N}}{\partial q_2} & \cdots & \frac{\partial x_{2N}}{\partial q_N} & \frac{\partial x_{2N}}{\partial p_1} & \frac{\partial x_{2N}}{\partial p_2} & \cdots & \frac{\partial x_{2N}}{\partial p_N} \end{pmatrix}, \quad (7.12)$$

we get:

$$\left(\frac{\partial F}{\partial q_1}, \ldots, \frac{\partial F}{\partial q_N}, \frac{\partial F}{\partial p_1}, \ldots, \frac{\partial F}{\partial p_N}\right) = \left(\frac{\partial F}{\partial x_1}, \frac{\partial F}{\partial x_2}, \ldots, \frac{\partial F}{\partial x_{2N}}\right) \left[\frac{\partial x}{\partial q} \quad \frac{\partial x}{\partial p}\right]. \quad (7.13)$$

Now, we can place (7.9), (7.13) in (7.2), getting:

$$\{F, G\} = \left(\frac{\partial F}{\partial x_1}, \frac{\partial F}{\partial x_2}, \ldots, \frac{\partial F}{\partial x_{2N}}\right) \left[\frac{\partial x}{\partial q} \quad \frac{\partial x}{\partial p}\right] [J] \begin{bmatrix} \frac{\partial x}{\partial q} \\ \frac{\partial x}{\partial p} \end{bmatrix} \begin{pmatrix} \frac{\partial G}{\partial x_1} \\ \frac{\partial G}{\partial x_2} \\ \vdots \\ \frac{\partial G}{\partial x_{2N}} \end{pmatrix}. \quad (7.14)$$

If we define the Poisson Matrix $[P]$ as:

$$[P] = \left[\frac{\partial x}{\partial q} \quad \frac{\partial x}{\partial p}\right] [J] \begin{bmatrix} \frac{\partial x}{\partial q} \\ \frac{\partial x}{\partial p} \end{bmatrix}, \quad (7.15)$$

we get:

$$\{F,G\} = \left(\frac{\partial F}{\partial x_1}, \frac{\partial F}{\partial x_2}, \ldots, \frac{\partial F}{\partial x_{2N}}\right)[P]\begin{pmatrix}\frac{\partial G}{\partial x_1}\\ \frac{\partial G}{\partial x_2}\\ \vdots \\ \frac{\partial G}{\partial x_{2N}}\end{pmatrix}. \tag{7.16}$$

The Poisson Matrix $[P]$ can be calculated using (7.15). However, the other way to calculate the matrix $[P]$ is to use the variables x_i and x_j, $i,j = 1,\ldots,2N,$ in place of F and G in (7.16). We will get vectors with a single 1 and multiple $0's$ as the entries in the vectors on both sides of the matrix $[P]$ in (7.16). Then the matrix multiplication on the right side will result in obtaining just one entry of the matrix $[P]$. Comparing both sides of (7.16) will then give:

$$\{F,G\} = \left(\frac{\partial F}{\partial x_1}, \frac{\partial F}{\partial x_2}, \ldots, \frac{\partial F}{\partial x_{2N}}\right)\begin{bmatrix}\{x_1,x_1\} & \{x_1,x_2\} & \cdots & \{x_1,x_{2N}\}\\ \{x_2,x_1\} & \vdots & & \vdots \\ \vdots & \vdots & & \vdots \\ \{x_{2N},x_1\} & \{x_{2N},x_1\} & \cdots & \{x_{2N},x_{2N}\}\end{bmatrix}\begin{pmatrix}\frac{\partial G}{\partial x_1}\\ \frac{\partial G}{\partial x_2}\\ \vdots \\ \frac{\partial G}{\partial x_{2N}}\end{pmatrix}, \tag{7.17}$$

which also means that:

$$[P] = \begin{bmatrix}\{x_1,x_1\} & \{x_1,x_2\} & \cdots & \{x_1,x_{2N}\}\\ \{x_2,x_1\} & \vdots & & \vdots \\ \vdots & \vdots & & \vdots \\ \{x_{2N},x_1\} & \{x_{2N},x_1\} & \cdots & \{x_{2N},x_{2N}\}\end{bmatrix}. \tag{7.18}$$

The formula (7.17) can be rewritten as:

$$\{F,G\} = \sum_{i=1}^{2N}\sum_{j=1}^{2N}\frac{\partial F}{\partial x_i}\cdot\{x_i,x_j\}\cdot\frac{\partial G}{\partial x_j}. \tag{7.19}$$

So, we have two ways to calculate the Poisson Brackets. We can either use the formulas (7.15) and (7.16), or we can use the formula (7.19). Both methods will work for any system of variables.

4. List of Basic Properties of Poisson Brackets for Systems with Constraints

Here are the results of all the calculations above, placed together for the convenience of the reader:

Assume $E, F,$ and G are some functions of $(q_1, q_2, ..., q_N, p_1, p_2, ..., p_N)$, and a and b are constants.

a) $\{F, G\} = -\{G, F\}$ (7.20)

b) $\{aF, G\} = a\{F, G\}$ (7.21)

c) $\{F, aG\} = a\{F, G\}$ (7.22)

d) $\{F + G, E\} = \{F, E\} + \{G, E\}$ (7.23)

e) $\{F, G + E\} = \{F, G\} + \{F, E\}$ (7.24)

f) $\{aF + bG, E\} = a\{F, E\} + b\{G, E\}$ (7.25)

h) $\{F, aG + bE\} = a\{F, G\} + b\{F, E\}$ (7.26)

i) $\{FG, E\} = F\{G, E\} + G\{F, E\}$ (7.27)

j) $\{F, GE\} = G\{F, E\} + E\{F, G\}$ (7.28)

h) Jacobi Identity (notice the cyclical nature of the rotation of functions)

$$\{\{F, G\}, E\} + \{\{G, E\}, F\} + \{\{E, F\}, G\} = 0 \quad (7.29)$$

i) If we change the variables from (q, p) to $(x_1, x_2, ..., x_{2N})$, where the new variables are given by:

$$x_1 = x_1(q, p),$$
$$x_2 = x_2(q, p),$$
$$\vdots$$
$$x_{2N} = x_{2N}(q, p),$$

then the Poisson Brackets expressed in new variables is given as:

$$\{F, G\} = \sum_{i=1}^{2N} \sum_{j=1}^{2N} \frac{\partial F}{\partial x_i} \cdot \{x_i, x_j\} \cdot \frac{\partial G}{\partial x_j}. \qquad (7.30)$$

Let us repeat here that the definition (7.1) of the Poisson Brackets used for constrained systems is identical to the definition used for non-constrained systems, and therefore, all properties above remain unchanged.

5. Poisson Brackets for Functions Equal on Constraints

Now, we will check how equality on constraints relates to the Poisson Brackets. The basic question is, "If functions are equal on constrains, are their Poisson Brackets also equal on constraints?"

Let us start with two functions $F(q_1, q_2, ..., q_N, p_1, p_2, ..., p_N)$ and $G(q_1, q_2, ..., q_N, p_1, p_2, ..., p_N)$. Assume that we have a set of constraints $\varphi_i(q_1, q_2, ..., q_N, p_1, p_2, ..., p_N) = 0$, $i = 1, ..., S$. We know from Chapter VI that all functions that are equal to F and G on constraints are given as:

$$F(q_1, q_2, ..., q_N, p_1, p_2, ..., p_N) + \sum_{i=1}^{S} u_i(q_1, q_2, ..., q_N, p_1, p_2, ..., p_N) \varphi_i(q_1, q_2, ..., q_N, p_1, p_2, ..., p_N)$$

and

$$G(q_1, q_2, ..., q_N, p_1, p_2, ..., p_N) + \sum_{j=1}^{S} w_j(q_1, q_2, ..., q_N, p_1, p_2, ..., p_N) \varphi_j(q_1, q_2, ..., q_N, p_1, p_2, ..., p_N)$$

To make it shorter, we will write respectively as $F + \sum_{i=1}^{S} u_i \varphi_i$ and $G + \sum_{j=1}^{S} w_j \varphi_j$. Let us now calculate the Poisson Brackets, using general rules from section 3. We get:

$$\left\{F + \sum_{i=1}^{S} u_i \varphi_i, G + \sum_{j=1}^{S} w_j \varphi_j\right\} = \left\{F, G + \sum_{j=1}^{S} w_j \varphi_j\right\} + \left\{\sum_{i=1}^{S} u_i \varphi_i, G + \sum_{j=1}^{S} w_j \varphi_j\right\} =$$

$$= \{F, G\} + \left\{F, \sum_{j=1}^{S} w_j \varphi_j\right\} + \left\{\sum_{i=1}^{S} u_i \varphi_i, G\right\} + \left\{\sum_{i=1}^{S} u_i \varphi_i, \sum_{j=1}^{S} w_j \varphi_j\right\} =$$

$$= \{F, G\} + \sum_{j=1}^{S} \{F, w_j \varphi_j\} + \sum_{i=1}^{S} \{u_i \varphi_i, G\} + \sum_{i,j=1}^{S} \{u_i \varphi_i, w_j \varphi_j\} =$$

$$= \{F, G\} + \sum_{j=1}^{S} \{F, w_j \varphi_j\} + \sum_{i=1}^{S} \{u_i \varphi_i, G\} + \sum_{i,j=1}^{S} \{u_i \varphi_i, w_j \varphi_j\} =$$

$$= \{F, G\} + \sum_{j=1}^{S} \left[w_j \{F, \varphi_j\} + \varphi_j \{F, w_j\} \right] +$$

$$+ \sum_{i=1}^{S} \left[u_i \{\varphi_i, G\} + \varphi_i \{u_i, G\} \right] + \sum_{i,j=1}^{S} \left[u_i \{\varphi_i, w_j \varphi_j\} + \varphi_i \{u_i, w_j \varphi_j\} \right] =$$

$$= \{F, G\} + \sum_{j=1}^{S} w_j \{F, \varphi_j\} + \sum_{j=1}^{S} \varphi_j \{F, w_j\} + \sum_{i=1}^{S} u_i \{\varphi_i, G\} + \sum_{i=1}^{S} \varphi_i \{u_i, G\} +$$

$$+ \sum_{i,j=1}^{S} \left[u_i w_j \{\varphi_i, \varphi_j\} + u_i \varphi_j \{\varphi_i, w_j\} + \varphi_i w_j \{u_i, \varphi_j\} + \varphi_i \varphi_j \{u_i, w_j\} \right].$$

So, we get:

$$\left\{F + \sum_{i=1}^{S} u_i \varphi_i, G + \sum_{j=1}^{S} w_j \varphi_j\right\} = \{F, G\} + \sum_{j=1}^{S} w_j \{F, \varphi_j\} + \sum_{j=1}^{S} \varphi_j \{F, w_j\} +$$

$$+ \sum_{i=1}^{S} u_i \{\varphi_i, G\} + \sum_{i=1}^{S} \varphi_i \{u_i, G\} + \tag{7.31}$$

$$+ \sum_{i,j=1}^{S} \left[u_i w_j \{\varphi_i, \varphi_j\} + u_i \varphi_j \{\varphi_i, w_j\} + \varphi_i w_j \{u_i, \varphi_j\} + \varphi_i \varphi_j \{u_i, w_j\} \right]$$

If we use the fact that constraints are equal to zeros on constraints, we get:

$$\left\{F+\sum_{i=1}^{S}u_i\varphi_i, G+\sum_{j=1}^{S}w_j\varphi_j\right\} = \{F,G\} + \sum_{j=1}^{S}w_j\{F,\varphi_j\} + \sum_{i=1}^{S}u_i\{\varphi_i,G\} +$$
$$+\sum_{i,j=1}^{S}u_iw_j\{\varphi_i,\varphi_j\} \qquad (7.32)$$

If we decide to only modify the first function, then all w_j's in (7.32) are equal to zero, and we get, on constraints:

$$\left\{F+\sum_{i=1}^{S}u_i\varphi_i, G\right\} = \{F,G\} + \sum_{i=1}^{S}u_i\{\varphi_i,G\} \qquad (7.33)$$

If we decide to only modify the second function, then all u_j's in the (7.32) are equal to zero, and we get, on constraints:

$$\left\{F, G+\sum_{j=1}^{S}w_j\varphi_j\right\} = \{F,G\} + \sum_{j=1}^{S}w_j\{F,\varphi_j\}. \qquad (7.34)$$

The conclusion from (7.32), (7.33), and (7.34) is that, in general, the Poisson Brackets will change if we modify a function in it, even if we use constraints to simplify the results. So, we will need to be careful when using constraints to simplify expressions, if we also use the Poisson Brackets.

6. Hamilton's Equations Using the Poisson Brackets for Functions Equal on Constraints

We have not defined the Hamiltonian for a general constrained system yet; we only did it for the example presented in chapter IV. Still, it is convenient to consider how such a Hamiltonian should work for systems with constraints.

Before defining the Hamiltonian in a general case of a constrained Lagrangian, we will describe how we expect such Hamiltonian to work. Namely, we expect that Hamilton's equations of motion (3.14), which we obtained for the hyperregular Lagrangian in Chapter III, will still work, with constraints added to them. In other words, we expect that the equations of motion (3.21) obtained in Chapter III, will be expressed in the form:

$$\dot{F} = \{F, H\},$$
$$\varphi_s(q, p) = 0, \qquad s = 1, \ldots, S. \qquad (7.35)$$

Notice that since the constraints were designed to hold in time, then the time derivatives of constraints must be equal to zeros on constraints. Then we must have:

$$\dot{\varphi}_s = \{\varphi_s, H\} = \sum_{j=1}^{S} w_j \varphi_j, \qquad s = 1, \ldots, S, \qquad (7.36)$$

since the time derivatives of constraints must be equal to zeros on constraints. Therefore they must be expressible by linear combinations of constraints, where w_j's are some functions of $(q_1, q_2, \ldots, q_N, p_1, p_2, \ldots, p_N)$. So, on constraints, we have:

$$\dot{\varphi}_s = \{\varphi_s, H\} = 0 \qquad s = 1, \ldots, S. \qquad (7.37)$$

Let us now consider replacing the function F with a function equal to it on constraints, namely

$$F + \sum_{j=1}^{S} w_j \varphi_j.$$

We will start with replacing it on the left side of the equation (7.35). We get:

$$\frac{d}{dt}\left(F + \sum_{j=1}^{S} w_j \varphi_j\right) = \{F, H\},$$
$$\frac{dF}{dt} + \frac{d}{dt}\left(\sum_{j=1}^{S} w_j \varphi_j\right) = \{F, H\}, \qquad (7.38)$$
$$\frac{dF}{dt} + \sum_{j=1}^{S} \frac{dw_j}{dt} \cdot \varphi_j + \sum_{j=1}^{S} w_j \cdot \frac{d\varphi_j}{dt} = \{F, H\}.$$

Since both φ_j and $\frac{d\varphi_j}{dt}$ are equal to zero on constraints, the left side simplifies to just $\frac{dF}{dt}$. This means that if we replace *F* on the left side with a function equal to it on constraints, we get an equivalent equation of motion. So, such replacements, which include substitutions and simplifications using constraints, are in general allowed.

Let us now look at the right side of the equations of motion (7.35). Let us replace *F* inside of the Poisson Brackets by $F + \sum_{j=1}^{S} w_j \varphi_j$. Using properties from section 3 of this chapter we get:

$$\left\{ F + \sum_{j=1}^{S} w_j \varphi_j, H \right\} = \{F, H\} + \left\{ \sum_{j=1}^{S} w_j \varphi_j, H \right\} =$$
$$= \{F, H\} + \sum_{j=1}^{S} \{w_j \varphi_j, H\} = \{F, H\} + \sum_{j=1}^{S} \left[w_j \{\varphi_j, H\} + \varphi_j \{w_j, H\} \right].$$
(7.39)

Since on constraints we have $\varphi_j = 0$, and (7.37) shows that also $\{\varphi_j, H\} = 0$, $j = 1, ..., S$, on constraints, then (7.39) shows that, on constraints, we have:

$$\left\{ F + \sum_{j=1}^{S} w_j \varphi_j, H \right\} = \{F, H\}.$$
(7.40)

This means, that since the equations of motion are supposed to work only on constraints, we may use the replacements in *F* on the right side of the Hamilton equations of motion (7.35). Combining (7.40) with the replacements on the left side of (7.35) that we studied earlier, we conclude that we are allowed to use replacements on both sides of (7.35). Notice that the replacements do not have to be identical on both sides; they are independent.

This property of the Hamilton's equations of motion is very useful when simplifying an *F* using constraints.

Finally, we need to look at the possibility of simplifying or modifying the Hamiltonian using constraints. In this case, we would replace the Hamiltonian *H* by $H + \sum_{j=1}^{S} w_j \varphi_j$. The right side of the Hamilton's equations (7.35) would then become:

$$\left\{F, H + \sum_{j=1}^{S} w_j \varphi_j\right\}$$

Simplifying this using the properties from section 4 of this chapter, we get:

$$\left\{F, H + \sum_{j=1}^{S} w_j \varphi_j\right\} = \{F, H\} + \left\{F, \sum_{j=1}^{S} w_j \varphi_j\right\} =$$

$$= \{F, H\} + \sum_{j=1}^{S} \{F, w_j \varphi_j\} = \{F, H\} + \sum_{j=1}^{S} \left[w_j \{F, \varphi_j\} + \varphi_j \{F, w_j\}\right].$$

Since the φ_j's are equal to zero on constraints, we get, on constraints:

$$\left\{F, H + \sum_{j=1}^{S} w_j \varphi_j\right\} = \{F, H\} + \sum_{j=1}^{S} w_j \{F, \varphi_j\}. \qquad (7.41)$$

Since we did not get just $\{F, H\}$ by itself on the right side of (7.41), we can see that, in general, we are not allowed to replace a Hamiltonian by another function that is equal to the Hamiltonian on constraints. If we do this, then in general we will not get the correct Hamilton's equations of motion, assuming that by "correct" we understand equations equivalent to the original Euler-Lagrange equations of motion obtained from the original Lagrangian. In this textbook, we will always require that the Hamilton's equations of motion (7.35) are equivalent to the Euler-Lagrange equations obtained from the original Lagrangian.

So, the Hamiltonian cannot be freely simplified by using constraints. Once we have a Hamiltonian that works in the above sense, we have to be very careful when trying to simplify it using constraints.

This result, while disappointing on some level, also gives us hope that if the Hamiltonian obtained for a constrained system by the traditional method taken from non-constrained systems is not working properly, then possibly this Hamiltonian may be modified by using constraints, until it does. By "working properly" we mean that it reproduces the original Euler-Lagrange equations of motion.

An example of this hope being realized was already given in chapter IV. In the example given there, we first got the Hamiltonian (4.27), which was not reproducing the correct Euler-Lagrange equations (4.23). However, then we used some substitutions using the constraints on the original Hamiltonian, and we got two Hamiltonians, (4.30) and (4.35), which both worked correctly by reproducing the correct equations of motion (4.23).

There is also another solution, possibly a more elegant, but less general one, namely defining new dynamical brackets that will replace the Poisson Brackets. The Hamilton's equations expressed by these new brackets, called "Dirac Brackets", will be discussed in detail in chapters XIII, XIV, and XV.

CHAPTER VIII

RETRIVING A FUNCTION FROM ITS PARTIAL DERIVATIVES

In this chapter, we want to show how to obtain a function from its partial derivatives. This material is a well-known part of regular calculus, and we present it here just for completeness. It will be used in this book in one place only; namely in Chapter IX when showing that the Hamiltonian of a constrained system is a function of positions and momenta only, and velocities can be eliminated.

1. Retrieving a Function from its Partial Derivatives

Let us assume that we have a differentiable function $f = f(x_1, x_2, ..., x_N)$, with partial derivatives denoted by $\frac{\partial f}{\partial x_1}, \frac{\partial f}{\partial x_2}, ..., \frac{\partial f}{\partial x_N}$. We will introduce the differential df of the function f as:

$$df = \frac{\partial f}{\partial x_1} dx_1 + \frac{\partial f}{\partial x_2} dx_2 + ... + \frac{\partial f}{\partial x_N} dx_N. \tag{8.1}$$

We will not attempt to precisely define the meaning of the differential (8.1); interested readers should look for books on differential forms. We will interpret the formula (8.1) as the statement that when we move from the point $(x_1, x_2, ..., x_N)$ to an infinitesimally close point $(x_1 + dx_1, x_2 + dx_2, ..., x_N + dx_N)$, then the infinitesimal change df in the value of the function f is given by the formula (8.1).

Then assume we have a path parametrized by a variable t, $a \leq t \leq b$, given as:

$$\begin{aligned} x_1 &= x_1(t), \\ x_2 &= x_2(t), \\ &\vdots \\ x_N &= x_N(t). \end{aligned} \tag{8.2}$$

Using the formula analogous to (8.1) for each variable in (8.2), we get the differentials:

$$dx_1 = \frac{dx_1}{dt}dt,$$

$$dx_2 = \frac{dx_2}{dt}dt,$$

$$\vdots \qquad (8.3)$$

$$dx_N = \frac{dx_N}{dt}dt,$$

which gives us infinitesimally small changes of variables, associated with infinitesimally small changes of the parameter t along the path we have chosen.

If we start with the value of the function f at the point $(x_1(a), x_2(a), ..., x_N(a))$, which is the beginning of the path, then we can calculate the value of the function f at the point $(x_1(b), x_2(b), ..., x_N(b))$ at the end of the path. This can be done by adding all changes df, along the path. Adding all infinitesimal changes is what we call an integral. So, we get:

$$f(x_1(b), x_2(b), ..., x_N(b)) = f(x_1(a), x_2(a), ..., x_N(a)) + \sum_{all\ df's} df.$$

Using the concept of an integral along the path, we can rewrite it as:

$$f(x_1(b), x_2(b), ..., x_N(b)) = f(x_1(a), x_2(a), ..., x_N(a)) + \int_{\substack{along \\ the\ path}} df. \qquad (8.4)$$

Using (8.1), we get

$$f(x_1(b), x_2(b), ..., x_N(b)) = f(x_1(a), x_2(a), ..., x_N(a)) +$$

$$+ \int_{\substack{along \\ the\ path}} \left[\frac{\partial f}{\partial x_1} dx_1 + \frac{\partial f}{\partial x_2} dx_2 + ... + \frac{\partial f}{\partial x_N} dx_N \right].$$

Using (8.3), we get:

$$f(x_1(b), x_2(b), ..., x_N(b)) = f(x_1(a), x_2(a), ..., x_N(a)) +$$

$$+ \int_{\text{along the path}} \left[\frac{\partial f}{\partial x_1} \cdot \frac{dx_1}{dt} dt + \frac{\partial f}{\partial x_2} \cdot \frac{dx_2}{dt} dt + ... + \frac{\partial f}{\partial x_N} \cdot \frac{dx_N}{dt} dt \right].$$

Finally, by factoring dt from the brackets and specifying the limits of the integral, we get the formula that can be effectively used, namely:

$$f(x_1(b), x_2(b), ..., x_N(b)) = f(x_1(a), x_2(a), ..., x_N(a)) +$$

$$+ \int_{t=a}^{t=b} \left[\frac{\partial f}{\partial x_1} \cdot \frac{dx_1}{dt} + \frac{\partial f}{\partial x_2} \cdot \frac{dx_2}{dt} + ... + \frac{\partial f}{\partial x_N} \cdot \frac{dx_N}{dt} \right] dt. \tag{8.5}$$

It may look strange that we would like to start with a function, calculate its partial derivatives, and then obtain the same function back. It seems like a futile exercise, and indeed, in this setting, it is. So, we do not use (8.5) to calculate the function f in specific cases. However, we will see in the next chapter that the analysis of the variables that show up in the derivatives and integrals in formula (8.5) allows us to prove in a general case that the Hamiltonian, while defined by using positions, momenta, and velocities, can be expressed by positions and momenta only.

2. An Example

Consider the function of two variables:

$$f(x, y) = xy^3 \tag{8.6}$$

It's partial derivatives are:

$$\frac{\partial f}{\partial x} = y^3,$$
$$\frac{\partial f}{\partial y} = 3xy^2. \tag{8.7}$$

Say, we want to start at $x = 0$, $y = 0$ and we want to calculate the original function (8.6) at the general point $x = A, y = B$.

We start with creating a path from $x = 0, y = 0$ to $x = A, y = B$. There exist infinitely many possibilities, but let us use a very simple one, namely:

$$x(t) = At,$$
$$y(t) = Bt, \qquad (8.8)$$
$$0 \leq t \leq 1.$$

It is obvious that for $t = 0$ we get the starting point $x = 0, y = 0,$ and for $t = 1$ we get the ending point $x = A, y = B,$ as required.

Using (8.8) and (8.3), we get the differentials:

$$dx = \frac{dx}{dt} dt = A dt,$$
$$dy = \frac{dy}{dt} dt = B dt. \qquad (8.9)$$

Placing (8.7), (8.8), and (8.9) into the formula (8.5), we get:

$$f(A,B) = f(x(1), y(1)) = f(x(0), y(0)) + \int_{t=0}^{t=1} \left[\frac{\partial f}{\partial x} \cdot \frac{dx}{dt} + \frac{\partial f}{\partial y} \cdot \frac{dy}{dt} \right] dt =$$

$$= f(0,0) + \int_{t=0}^{t=1} \left[(y^3) \cdot A + (3xy^2) \cdot B \right] dt =$$

$$= 0 + \int_{t=0}^{t=1} \left[((Bt)^3) \cdot A + (3(At)(Bt)^2) \cdot B \right] dt =$$

$$= \int_{t=0}^{t=1} \left[AB^3 t^3 + 3AB^3 t^3 \right] dt = \int_{t=0}^{t=1} \left[4AB^3 t^3 \right] dt =$$

$$= 4AB^3 \int_{t=0}^{t=1} t^3 dt = 4AB^3 \cdot \left[\frac{t^4}{4} \right]_{t=0}^{t=1} = 4AB^3 \cdot \left[\frac{1}{4} - \frac{0}{4} \right] = AB^3$$

So, we obtained:

$$f(A,B) = AB^3$$

Since A and B can represent any numbers, we can replace them by x and y, getting:

$$f(x,y) = xy^3.$$

This result is not surprising, since it just reproduces the function (8.6) with which we started.

The reader can verify that combination of two separate paths is possible and will give the same result. For example, we could use the paths:

$$\begin{aligned} x(t) &= At, \\ y(t) &= 0, \\ 0 &\leq t \leq 1, \end{aligned} \tag{8.10}$$

and

$$\begin{aligned} x(t) &= A, \\ y(t) &= Bt, \\ 0 &\leq t \leq 1. \end{aligned} \tag{8.11}$$

The first path will take us from the point $x = 0$, $y = 0$ to the point $x = A$, $y = 0$. The second path will take us from the point $x = A$, $y = 0$ to the point $x = A$, $y = B$.. So, we would first get $f(A,0)$ and then, starting with $f(A,0)$, we separately get $f(A,B)$. We are going to skip the details since they are very similar to the previous calculation.

It is important to notice that since we started with a specific, existing function, then no matter what path we choose, we will get back the same function with which we started.

3. A Comment

A less experienced reader may conclude that we can start with arbitrary formulas for the partial derivatives $\frac{\partial f}{\partial x_1}, \frac{\partial f}{\partial x_2}, ..., \frac{\partial f}{\partial x_N}$, and then the formula (8.5) will give us the function that has these partial derivatives. It is important to realize that this is not the case. In general, the formula (8.5) will give different results when we choose different paths, so that no single function will be defined.

In basic calculus, it can be shown that the condition for getting a uniquely defined function by the process presented in this chapter is that mixed second derivatives are equal on a certain part of the domain. However, in Volume I of this book we avoid checking the mixed second derivatives, because we use the formula (8.5) by starting with an existing function, calculating its partial derivatives, and then getting the same function back. Since we have shown above that the formula (8.5), if obtained from an existing function, gives the same function back, there is no need to check the uniqueness by studying the mixed derivatives. The need to check the equality of the mixed second derivatives will first appear in Volume III.

CHAPTER IX

THE PRE-HAMILTONIAN FOR A CONSTRAINED LAGRANGIAN

In this chapter, we want to concentrate on constructing a pre-Hamiltonian for a given constrained Lagrangian. We may use formulas from previous chapters while giving them new numbers, so they will be easier to refer to in this chapter.

1. Introduction of the pre-Hamiltonian for a Constrained Lagrangian

As before, let us assume that we have N-dimensional mechanical system, described by the position-velocity variables $(q_1, q_2, ..., q_N, v_1, v_2, ..., v_N)$, and a constrained Lagrangian:

$$L = L(q, v) = L(q_1, q_2, ..., q_N, v_1, v_2, ..., v_N). \tag{9.1}$$

Then, we extend our space to include the momenta. To make it shorter, we may replace the notation $(q_1, q_2, ..., q_N, v_1, v_2, ..., v_N, p_1, p_2, ..., p_N)$ by (q, v, p). At the beginning of our analysis, we consider all these variables to be completely independent of each other.

Now we define the pre-Hamiltonian as:

$$K = K(q, v, p) = \sum_{i=1}^{N} p_i \cdot v_i - L(q, v), \tag{9.2}$$

In (9.2) the variables (q, v, p) are treated as independent.

Notice that we defined the pre-Hamiltonian by almost the same formula as the Hamiltonian was defined by the formula (3.6) in Chapter III. The difference is that in Chapter III we had a non-constrained system, for which, by definition, all velocities could be expressed by positions and momenta by inverting the Legendre Transformations (3.1). Consequently, the Hamiltonian defined by formula (3.1) was defined as a function of positions and momenta only. Here, in the case of a constrained system, not all velocities can be expressed by positions and momenta using the Legendre Transformations only. The very definition of the constrained system is that this cannot be done. So, we need to analyze it further, and we begin with treating positions, velocities, and momenta on the same footing, as independent variables. At this point in our analysis of the constrained system we assume that we do not have the relations given by (3.1) yet.

Calculating the partial derivatives of the pre-Hamiltonian (9.2), still assuming independence of (q,v,p), we get:

$$\frac{\partial K}{\partial q_i} = -\frac{\partial L}{\partial q_i},$$
$$\frac{\partial K}{\partial v_i} = p_i - \frac{\partial L}{\partial v_i},$$
$$\frac{\partial K}{\partial p_i} = v_i.$$
(9.3)

Continuing the use of the assumption of the independence of the variables (q,v,p), we define the differential of the pre-Hamiltonian (9.2), as:

$$dK = \sum_{i=1}^{N} \frac{\partial K}{\partial q_i} \cdot dq_i + \sum_{i=1}^{N} \frac{\partial K}{\partial v_i} \cdot dv_i + \sum_{i=1}^{N} \frac{\partial K}{\partial p_i} \cdot dp_i.$$
(9.4)

Substituting the partial derivatives (9.3) into (9.4), we obtain:

$$dK = \sum_{i=1}^{N} \left(-\frac{\partial L}{\partial q_i}\right) \cdot dq_i + \sum_{i=1}^{N} \left(p_i - \frac{\partial L}{\partial v_i}\right) \cdot dv_i + \sum_{i=1}^{N} v_i \cdot dp_i.$$
(9.5)

Let us stress again that the formula (9.5) was obtained with the assumption that the variables (q,v,p) are independent of each other.

Only now we introduce the Legendre Transformations for the constrained Lagrangian (9.1). We define it by the same formulas (3.1) as we did for the hyperregular case in Chapter III, namely:

$$p_i = \frac{\partial L}{\partial v_i}, \quad i = 1, \ldots, N. \tag{9.6}$$

However, here we do not look at the Legendre Transformations as a change of variables, like we did in Chapter III. Instead, we consider it to be a set of constraints on the space given by variables (q, v, p). These constraints define a subspace of the space given by (q, v, p), the subspace made by such points that satisfy the Legendre Transformations equations (9.6).

Notice that since different equations (9.6) have different momenta on the left side, and only positions and velocities on the right side, then they are independent equations in the sense that none of them can be obtained as a combination of the remaining ones. This means that the constraints (9.6) reduce the dimension of the original (q, v, p) space from $3N$ to $2N$.

2. A Simple Example

Notice that the pre-Hamiltonian given by (9.2) is a function of (q, v, p). We will need the pre-Hamiltonian to be a function of (q, p) only. In this section, we will show an example that explains what we mean by this.

Showing that the Hamiltonian is a function of (q, p) only was easy in the case of the hyperregular Lagrangian, covered in Chapter III. The Legendre Transformations presented there could be solved for all velocities, and then the solutions could be used to remove the velocities from the formula for the Hamiltonian. Even more, any function of (q, v, p) could be easily expressed as the function of (q, p) in Chapter III, for the points satisfying the Legendre Transformations.

In the case of the constrained Lagrangians, it is not as simple. By definition, in the constrained case, we cannot solve the Legendre Transformations for velocities to eliminate them. Therefore, not all functions of (q, v, p) can be rewritten to be functions of (q, p) only, even for the points satisfying the Legendre Transformations. However, because of the specific form of the pre-

Hamiltonian (9.2), it can be shown that that this pre-Hamiltonian can be expressed by the (q,p) only, for the points that satisfy the Legendre Transformations.

To explain better what we mean, let us look at a simple example. Say we have a function of three variables (x,y,z), given as:

$$f(x,y,z) = x^3 + 2x + 2y - 5z. \tag{9.7}$$

Assume also that we have a constraint given by:

$$\varphi(x,y,z) = x + 2y - 5z = 0. \tag{9.8}$$

We can rewrite $f(x,y,z)$ as:

$$f(x,y,z) = x^3 + x + x + 2y - 5z, \tag{9.9}$$

and then, using the constraint (9.8), we can replace the last part of the function in (9.9) by zero, getting a different function, namely:

$$g(x,y,z) = x^3 + x. \tag{9.10}$$

Then, we have

$$f(x,y,z) = g(x,y,z), \tag{9.11}$$

for every point (x,y,z) that satisfies the constraint (9.8). However, at the same time, we can define another function, namely:

$$h(x) = x^3 + x. \tag{9.12}$$

Now we can claim that the function $h(x)$ from (9.12) is equal to the function $f(x,y,z)$ from (9.7) on the constraint (9.8). Mathematically speaking, they are not equal as functions, since they operate on different domains. However, they are equal in the sense that just knowing the value of the variable x of a point allows us to get the value of both functions (on constraints), and these values are identical.

So, on the constraints (9.8), it makes sense to replace $f(x,y,z) = x^3 + 2x + 2y - 5z$ by just $f(x) = x^3 + x$. This is what we mean by the statement that the function $f(x,y,z)$ is the function of just x, which we can denote by $f(x)$.

Notice that the statement that $f(x,y,z)$ is a function of just x on the constraint is due to a specific formula for $f(x,y,z)$, which allows for the elimination of other variables on the constraint (9.8). This will not always be true for other functions of (x,y,z).

It is worth noting that, using the results of Chapter VI, we know that there exists such a function $u(x,y,z)$ that we have:

$$f(x,y,z) = f(x) + u(x,y,z) \cdot \varphi(x,y,z), \tag{9.13}$$

for all points (x,y,z), including the points that do not satisfy the constraint (9.8). We could easily calculate such $u(x,y,z)$ from (9.13), but we have no use for this calculation here.

3. The pre-Hamiltonian as a Function of the Positions and Momenta Only

After looking at the example, we now go back to the pre-Hamiltonian (9.2). The pre-Hamiltonian is a function of positions, momenta, and velocities. In this section, we want to show that when the pre-Hamiltonian (9.2) is restricted to only such points (q,v,p) that satisfy the Legendre Transformations constraints (9.6), then it becomes the function of (q,p) only, in the sense explained in the previous section.

We will reconstruct the pre-Hamiltonian (9.2), using its differential (9.5). Let us start with one fixed single point in the (q,v,p) space, assuming that this point satisfies the Legendre Transformations (9.6). For our considerations, it does not matter with which specific point we start. So, say we denote this starting point by (q_0, v_0, p_0).

Earlier in Chapter V, formula (5.4), we re-wrote the Legendre Transformations constraints (9.6), as:

$$v_k = v_k(q, p, v_{M+1}, v_{M+2},..., v_N), \qquad k = 1,..., M < N,$$
$$\varphi_s(q, p) = 0, \qquad s = M+1,..., N. \tag{9.14}$$

This form of the Legendre Transformations separates all velocities into the first M velocities that are dependent on the positions, momenta, and the remaining $N - M$ velocities. The last $N - M$ velocities can be thought of as independent. This division is somewhat arbitrary; different velocities can be thought of as dependent or independent. Let us assume that we made this arbitrary choice, and that we keep it.

Using the above, we will split the velocities into two sets, dependent and independent, denoted respectively by (v_D, v_I).

In what follows it is convenient to explicitly show that v_D is dependent, so we describe it by $v_D = v_D(q, v_I, p)$. Then, the general point that satisfies the Legendre Transformations (9.14), will be denoted by $(q, v_D(q, v_I, p), v_I, p)$. Notice that, if we want this point to satisfy the Legendre Transformations (9.14), the positions and momenta are also not independent. However, it is not beneficial to include this fact into the notation, so we just remember this and not include it into the notation.

We assume that the initial point described above satisfies both lines of the Legendre Transformations (9.14). Therefore, our initial point (q_0, v_0, p_0) can be denoted as $(q_0, v_D(q_0, v_{I0}, p_0), v_{I0}, p_0)$.

Then, we take another arbitrary point satisfying the Legendre Transformations. We will call it a terminal point and denote it by $(q_T, v_D(q_T, v_{IT}, p_T), v_{IT}, p_T)$, where T in the subscript stands for "terminal." Again, by assuming that the entire point satisfies the Legendre Transformations, we automatically assume that this point satisfies both lines in the Legendre Transformations (9.14).

Before going from the initial to the terminal point, let us define the "transition" point. This is the point where the positions and momenta are taken from the terminal point, but the independent velocities are taken from the initial point. This point is then given as $(q_T, v_D(q_T, v_{I0}, p_T), v_{I0}, p_T)$. Notice that since the dependent velocities are given by the formulas taken from the first line of the Legendre Transformations, and the positions and velocities are taken from the terminal point, then this transition point also satisfies the Legendre Transformations (9.14). We will consider this point as being somewhere "on the road" for our path from the initial to the terminal point.

The pre-Hamiltonian (9.2) at the "transition" point will be given by:

$$K = K(q_T, v_D(q_T, v_{I0}, p_T), v_{I0}, p_T). \tag{9.15}$$

Notice that the pre-Hamiltonian (9.15) at the transition point can be considered to be a function of (q_T, p_T) only, because the velocities in (9.15) are still the same initial velocities from the initial point $(q_0, v_D(q_0, v_{I0}, p_0), v_{I0}, p_0)$. So, these velocities are constants, not variables.

Now we will use (9.15) and the differential dK given in (9.4) and (9.5) to calculate the pre-Hamiltonian at the terminal point. Let us start to define a path from the "transition" point to the terminal point. The simplest path we can think of is parametrized by t, $0 \leq t \leq 1$:

$q_1 = q_1(t) = (q_T)_1,$
$q_2 = q_2(t) = (q_T)_2,$
$\vdots$
$q_N = q_N(t) = (q_T)_N,$
$v_{D1} = v_{D1}(t) = v_{D1}(q(t), v_I(t), p(t)),$

(continued on the next page)

$$v_{D2} = v_{D2}(t) = v_{D2}(q(t), v_I(t), p(t)),$$
$$\vdots$$
$$v_{DM} = v_{DM}(t) = v_{DM}(q(t), v_I(t), p(t)),$$
$$v_{I1} = v_{I1}(t) = v_{I1} + (v_{T1} - v_{I1}) \cdot t,$$
$$v_{I2} = v_{I2}(t) = v_{I2} + (v_{T2} - v_{I2}) \cdot t,$$
$$\vdots \qquad (9.16)$$
$$v_{I(N-M)} = v_{I(N-M)}(t) = v_{I(N-M)} + (v_{T(N-M)} - v_{I(N-M)}) \cdot t,$$
$$p_1 = p_1(t) = (p_T)_1,$$
$$p_2 = p_2(t) = (p_T)_2,$$
$$\vdots$$
$$p_N = p_N(t) = (p_T)_N.$$

Notice that the path (9.16) gives us the 'transition" point for $t=0$, and the terminal point for $t=1$. Notice also that positions and momenta are constant along that path.

The differentials obtained from the path (9.16) are:

$$dq_1 = 0 \cdot dt,$$
$$dq_2 = 0 \cdot dt,$$
$$\vdots$$
$$dq_N = 0 \cdot dt,$$
$$dv_{D1} = \frac{d}{dt}\left[v_{D1}(q(t), v_I(t), p(t))\right] \cdot dt,$$
$$dv_{D2} = \frac{d}{dt}\left[v_{D2}(q(t), v_I(t), p(t))\right] \cdot dt,$$
$$\vdots$$
$$dv_{DM} = \frac{d}{dt}\left[v_{D2}(q(t), v_I(t), p(t))\right] \cdot dt,$$
$$dv_{I1} = (v_{T1} - v_{I1}) \cdot dt,$$
$$dv_{I2} = (v_{T2} - v_{I2}) \cdot dt,$$
$$\vdots$$
$$dv_{I(N-M)} = (v_{T(N-M)} - v_{I(N-M)}) \cdot dt,$$

(continued on the next page)

$$dp_1 = 0 \cdot dt,$$
$$dp_2 = 0 \cdot dt,$$
$$\vdots \qquad (9.17)$$
$$dp_N = 0 \cdot dt.$$

Notice that every point along the path satisfies the Legendre Transformations (9.14) since the positions and momenta are constant, and the formulas from (9.14) give the dependent velocities.

Now we use the formula (8.4) from Chapter VIII and the formula (9.15) to calculate the pre-Hamiltonian at the terminal point. We get:

$$K\left(q_T, v_D\left(q_T, v_{IT}, p_T\right), v_{IT}, p_T\right) = K\left(q_T, v_D\left(q_T, v_{I0}, p_T\right), v_{I0}, p_T\right) + \int\limits_{\substack{along \\ the\ path}} dK. \qquad (9.18)$$

Using the formula (9.4) for dK in (9.18) we get:

$$K\left(q_T, v_D\left(q_T, v_{IT}, p_T\right), v_{IT}, p_T\right) = K\left(q_T, v_D\left(q_T, v_{I0}, p_T\right), v_{I0}, p_T\right) +$$
$$+ \int\limits_{\substack{along \\ the\ path}} \left[\sum_{i=1}^{N} \frac{\partial K}{\partial q_i} \cdot dq_i + \sum_{i=1}^{N} \frac{\partial K}{\partial v_i} \cdot dv_i + \sum_{i=1}^{N} \frac{\partial K}{\partial p_i} \cdot dp_i \right]. \qquad (9.19)$$

Formula (9.17) tells us that $dq_i = 0 \cdot dt$, and $dp_i = 0 \cdot dt$, for $i = 1,...,N$. So we get:

$$K\left(q_T, v_D\left(q_T, v_{IT}, p_T\right), v_{IT}, p_T\right) = K\left(q_T, v_D\left(q_T, v_{I0}, p_T\right), v_{I0}, p_T\right) +$$
$$+ \int\limits_{\substack{along \\ the\ path}} \left[\sum_{i=1}^{N} \frac{\partial K}{\partial q_i} \cdot 0 \cdot dt + \sum_{i=1}^{N} \frac{\partial K}{\partial v_i} \cdot dv_i + \sum_{i=1}^{N} \frac{\partial K}{\partial p_i} \cdot 0 \cdot dt \right].$$

Simplifying the above, we get:

$$K\left(q_T, v_D\left(q_T, v_{IT}, p_T\right), v_{IT}, p_T\right) = K\left(q_T, v_D\left(q_T, v_{I0}, p_T\right), v_{I0}, p_T\right) +$$
$$+ \int\limits_{\substack{along \\ the\ path}} \left[\sum_{i=1}^{N} \frac{\partial K}{\partial v_i} \cdot dv_i\right]. \qquad (9.20)$$

Recall that the partial derivatives $\frac{\partial K}{\partial v_i}$ in formula (9.4) were obtained directly from (9.2), and they were calculated with the assumption that all positions, velocities, and momenta were treated as independent. This means that we still use the formula (9.3) to get the partial derivatives. Placing the result from (9.3) in (9.20) we get:

$$K\left(q_T, v_D\left(q_T, v_{IT}, p_T\right), v_{IT}, p_T\right) = K\left(q_T, v_D\left(q_T, v_{I0}, p_T\right), v_{I0}, p_T\right) +$$
$$+ \int\limits_{\substack{along \\ the\ path}} \left[\sum_{i=1}^{N} \left(p_i - \frac{\partial L}{\partial v_i}\right) \cdot dv_i\right]. \qquad (9.21)$$

The Legendre Transformations (9.14) can also be written in the form (9.6), which can be modified to:

$$p_i - \frac{\partial L}{\partial v_i} = 0, \ i = 1, ..., N. \qquad (9.22)$$

As we established earlier, each point along the path (9.16) satisfies the Legendre Transformations. So, we can substitute (9.22) into (9.21), getting:

$$K\left(q_T, v_D\left(q_T, v_{IT}, p_T\right), v_{IT}, p_T\right) = K\left(q_T, v_D\left(q_T, v_{I0}, p_T\right), v_{I0}, p_T\right) +$$
$$+ \int\limits_{\substack{along \\ the\ path}} \left[\sum_{i=1}^{N} (0) \cdot dv_i\right]. \qquad (9.23)$$

Therefore, we get:

$$K\left(q_T, v_D\left(q_T, v_{IT}, p_T\right), v_{IT}, p_T\right) = K\left(q_T, v_D\left(q_T, v_{I0}, p_T\right), v_{I0}, p_T\right). \qquad (9.24)$$

Recall that the terminal point $\left(q_T, v_D(q_T, v_{IT}, p_T), v_{IT}, p_T\right)$ was any point satisfying the Legendre Transformations. So, there is no reason to keep the subscript T there anymore (earlier we used it since it was a good practice not to have the same symbol inside of a definite integral and in expressions outside it). So, we get, for all points satisfying the Legendre Transformations:

$$K\left(q, v_D(q, v_I, p), v_I, p\right) = K\left(q, v_D(q, v_{I0}, p), v_{I0}, p\right). \tag{9.25}$$

Both sides look almost identical. However, notice that we have the independent variables v_I on the left side, and v_{I0}, which are not variables, but fixed numbers describing the fixed initial point, on the right side. So, the function K is not a function of the variables v_I. Also, since $v_D = v_D(q, v_{I0}, p)$ are not independent variables, we can drop them from the list of variables in the function. The conclusion is that the pre-Hamiltonian K, when restricted to the points satisfying the Legendre Transformations, is a function of (q, p) only. So, we now have:

$$K = K(q, p) = K\left(q, v_D(q, v_{I0}, p), v_{I0}, p\right). \tag{9.26}$$

Therefore, the pre-Hamiltonian K can be written as a function of (q, p) only.

4. Some Remarks

The formula obtained in (9.26) is valid only for the points satisfying the Legendre Transformations. For other points the definitions (9.26) and (9.2) would, in general, give different results.

Using the above integrals is not a very effective method of actually calculating a formula for the pre-Hamiltonian. While it can be used if we have no other choice, in the most practical application we manage to combine the definition of the pre-Hamiltonian (9.2) with the Legendre Transformations in the form (9.6) and (9.14) to eliminate all velocities from (9.2), and therefore to get the pre-Hamiltonian that is a function of (q, p) only.

Finally, a very important note: The fact that we got the pre-Hamiltonian K expressed as a function of (q, p) only, as in (9.26), does not mean that this pre-Hamiltonian will give us the correct equations of motion via typical differentiation, or via the Poisson Brackets, as did the Hamiltonian in the hyperregular case. Actually, it will turn out that, in general, the pre-Hamiltonian (9.26) will have to be modified to produce the correct equations of motion. The pre-Hamiltonian modified to give the correct equations of motion will be called the Hamiltonian. Obtaining the Hamiltonian from the pre-Hamiltonian is covered in Chapter XI.

CHAPTER X

ELEMENTS OF LINEAR ALGEBRA

We include this chapter to make the text more self-contained. It is not our goal to provide an introduction to linear algebra here; we assume that a reader of this book has basic knowledge of how vectors and matrices work. We just need one specific result that will be used in the next chapter.

1. An Important Theorem

The theorem below will be used in Chapter XI, Volume I. We provide a proof because it is not easily available in typical textbooks.

Theorem:

Assume that:

1) V is an n-dimensional vector space with a usual dot product.

2) We have a given finite set of vectors $\{\vec{w}_1, \vec{w}_2, ..., \vec{w}_k\}$ in V.

3) U is the set of all vectors $\vec{u}$ in V that satisfy

$$\vec{u} \cdot \vec{w}_1 = 0,$$
$$\vec{u} \cdot \vec{w}_2 = 0,$$
$$\vdots$$
$$\vec{u} \cdot \vec{w}_k = 0,$$

where the dot in the above means the dot product.

4) A vector $\vec{x}$ in V satisfies $\vec{x} \cdot \vec{u} = 0$ for every vector $\vec{u}$ from U.

With the above assumptions, we have the following theorem:

The vector $\vec{x}$ is a linear combination of the vectors $\vec{w}_1, \vec{w}_2, ..., \vec{w}_k$.

Notice that in the above, we do not assume that the vectors $\vec{w}_1, \vec{w}_2, ..., \vec{w}_k$ are linearly independent.

A reader not interested in the details of this linear algebra proof may skip it, accept the theorem above, and go to the next chapter.

Proof: Let us define the set W as the span of the vectors $\vec{w}_1, \vec{w}_2, ..., \vec{w}_k$. Then take a set of orthonormal vectors $\vec{e}_1, \vec{e}_2, ..., \vec{e}_m$ that have the same span. Notice that the number of orthonormal vectors $\vec{e}_1, \vec{e}_2, ..., \vec{e}_m$ does not have to be the same as the number of the original vectors $\vec{w}_1, \vec{w}_2, ..., \vec{w}_k$, because the vectors $\vec{w}_1, \vec{w}_2, ..., \vec{w}_k$ do not have to be linearly independent.

Then, since we can use either set to express any vector in that span, we have:

$$\vec{e}_i = \sum_{j=1}^{k} \alpha_j \vec{w}_j, \quad i=1,...,n, \quad \text{for some real numbers } \alpha_j, \text{ and}$$

$$\vec{w}_i = \sum_{j=1}^{m} \beta_j \vec{e}_j, \quad i=1,...,k, \quad \text{for some real numbers } \beta_j.$$

Let us now extend the set of orthonormal vectors $\vec{e}_1, \vec{e}_2, ..., \vec{e}_m$ to the complete orthonormal base of the vector space V. Let us denote this complete orthonormal base by

$$\vec{e}_1, \vec{e}_2, ..., \vec{e}_m, \vec{E}_{m+1}, \vec{E}_{m+2}, ..., \vec{E}_n.$$

In this base, any vector $\vec{u}$ from the set U can be written as:

$$\vec{u} = \sum_{i=1}^{m} \gamma_i \vec{e}_i + \sum_{i=m+1}^{n} \lambda_i \vec{E}_i.$$

Then we have, using the typical properties of orthonormal bases and the fact that $\vec{u} \cdot \vec{w}_i = 0, \quad i = 1,...,k,$:

$$\gamma_i = \vec{u} \cdot \vec{e}_i = \vec{u} \cdot \left(\sum_{j=1}^{k} \alpha_j \vec{w}_j \right) = \sum_{j=1}^{k} \alpha_j \left(\vec{u} \cdot \vec{w}_j \right) = \sum_{j=1}^{k} \alpha_j \cdot 0 = 0.$$

So

$$\gamma_i = 0.$$

So, the $\vec{u}$ satisfying $\vec{u} \cdot \vec{w}_i = 0, \quad i = 1,...,k,$ must be:

$$\vec{u} = \sum_{i=m+1}^{n} \lambda_i \vec{E}_i.$$

Now, take any vector $\vec{x} = \sum_{i=1}^{m} \varepsilon_i \vec{e}_i + \sum_{i=m+1}^{n} \eta_i \vec{E}_i$, such that $\vec{x} \cdot \vec{u} = 0$ for any $\vec{u}$ from the set U. We already know that any such $\vec{u}$ can be written as $\vec{u} = \sum_{i=m+1}^{n} \lambda_i \vec{E}_i$. We can choose $\vec{u} = \vec{E}_j$, and then we have:

$$0 = \vec{x} \cdot \vec{u} = \vec{x} \cdot \vec{E}_i = \eta_i.$$

So $\eta_i = 0$ and we have

$$\vec{x} = \sum_{i=1}^{m} \varepsilon_i \vec{e}_i + \sum_{i=m+1}^{n} 0 \cdot \vec{E}_i,$$

therefore

$$\vec{x} = \sum_{i=1}^{m} \varepsilon_i \vec{e}_i.$$

Then, since we have $\vec{e}_i = \sum_{j=1}^{k} \alpha_j \vec{w}_j$, we get:

$$\vec{x} = \sum_{i=1}^{m} \varepsilon_i \vec{e}_i = \sum_{i=1}^{m} \varepsilon_i \left(\sum_{j=1}^{k} \alpha_{ij} \vec{w}_j \right) = \sum_{j=1}^{k} \left(\sum_{i=1}^{m} \varepsilon_i \alpha_{ij} \right) \vec{w}_j.$$

Introducing the symbol $\kappa_j = \left(\sum_{i=1}^{m} \varepsilon_i \alpha_{ij} \right)$ we get:

$$\vec{x} = \sum_{j=1}^{k} \kappa_j \vec{w}_j.$$

Therefore, the vector $\vec{x}$ is a linear combination of the vectors $\vec{w}_1, \vec{w}_2, \ldots, \vec{w}_k$. Q.E.D.

To express the theorem above in words: it says that if we have a set A of vectors, and we define the set B as made of all vectors orthogonal to all vectors in A, then any vector which is orthogonal to all vectors in B can be written as a linear combination of the vectors from the starting set A.

CHAPTER XI

THE HAMILTONIAN FOR A REGULAR CONSTRAINED LAGRANGIAN

In this chapter we want to concentrate on constructing the final form of the Hamiltonian for a given regular constrained Lagrangian. We are going to start by recalling what we have established so far. We will give new numbers to equations and formulas that were introduced in earlier chapters, so they will be easier to refer to in this chapter.

This chapter is rather complicated, as we are trying to show as many details of the construction as possible. For a reader that is eager to skip the details, going directly to Chapter XII may be an option. Chapter XII recalls all results of this chapter, but without detailed explanations.

1. The Differential of the pre-Hamiltonian and the Differentials of Legendre Transformations

As before, let us assume that we have N-dimensional mechanical system, described by the position-velocity variables $(q_1, q_2, ..., q_N, v_1, v_2, ..., v_N)$, and a constrained Lagrangian:

$$L = L(q,v) = L(q_1, q_2, ..., q_N, v_1, v_2, ..., v_N). \tag{11.1}$$

We also have a pre-Hamiltonian K defined in (9.2) in Chapter IX as:

$$K = K(q,v,p) = \sum_{i=1}^{N} p_i \cdot v_i - L(q,v). \tag{11.2}$$

In (11.2), the variables (q, v, p) are treated as independent.

In Chapter IX, in formulas (9.4) and (9.5), we calculated the differential of the pre-Hamiltonian (11.2) as:

$$dK = \sum_{i=1}^{N} \frac{\partial K}{\partial q_i} \cdot dq_i + \sum_{i=1}^{N} \frac{\partial K}{\partial v_i} \cdot dv_i + \sum_{i=1}^{N} \frac{\partial K}{\partial p_i} \cdot dp_i. \tag{11.3}$$

Then we get:

$$dK = \sum_{i=1}^{N}\left(-\frac{\partial L}{\partial q_i}\right)\cdot dq_i + \sum_{i=1}^{N}\left(p_i - \frac{\partial L}{\partial v_i}\right)\cdot dv_i + \sum_{i=1}^{N} v_i \cdot dp_i. \tag{11.4}$$

The result (11.4) was obtained by treating (q, v, p) as a set of independent variables.

However, if we look at the result (11.4) only for the points satisfying the Legendre Transformations (9.6), we have $p_i - \frac{\partial L}{\partial v_i} = 0$, so we get:

$$dK = \sum_{i=1}^{N}\left(-\frac{\partial L}{\partial q_i}\right)\cdot dq_i + \sum_{i=1}^{N} 0 \cdot dv_i + \sum_{i=1}^{N} v_i \cdot dp_i, \text{ and then}$$

$$dK = \sum_{i=1}^{N}\left(-\frac{\partial L}{\partial q_i}\right)\cdot dq_i + \sum_{i=1}^{N} v_i \cdot dp_i. \tag{11.5}$$

Let us stress that the result (11.5) was obtained with the assumption that (q, v, p) are independent, and it holds only for points satisfying the Legendre Transformations.

Then, we have shown in Chapter IX that, when restricted to the points satisfying the Legendre Transformations, the pre-Hamiltonian (11.2) can be written as a function of just positions and momenta:

$$K = K(q, p). \tag{11.6}$$

Using (11.6), we can write for the differential of the pre-Hamiltonian:

$$dK = \sum_{i=1}^{N} \frac{\partial K}{\partial q_i} \cdot dq_i + \sum_{i=1}^{N} \frac{\partial K}{\partial p_i} \cdot dp_i. \tag{11.7}$$

Let us stress that the partial derivatives in (11.7) were obtained by treating (q,p) in (11.6) as independent, despite of possible constraints that may relate some of them in our constrained system.

As always, we need to remember that the partial derivatives in (11.7) and the partial derivatives in (11.3) are in general different, despite of the fact that we use the same symbols for them, because the function K is treated as a function of different variables for the two calculations.

Let us interpret dq_i and dp_i as infinitesimal changes of q_i and p_i. If we restrict ourselves to points satisfying the Legendre Transformations, then both the formulas (11.5) and (11.7) will give the same infinitesimal change dK of the pre-Hamiltonian K.

So, comparing (11.5) and (11.7), we get:

$$\sum_{i=1}^{N}\left(-\frac{\partial L}{\partial q_i}\right)\cdot dq_i + \sum_{i=1}^{N} v_i \cdot dp_i = \sum_{i=1}^{N}\frac{\partial K}{\partial q_i}\cdot dq_i + \sum_{i=1}^{N}\frac{\partial K}{\partial p_i}\cdot dp_i. \tag{11.8}$$

Subtracting the left side, we get:

$$\sum_{i=1}^{N}\frac{\partial K}{\partial q_i}\cdot dq_i + \sum_{i=1}^{N}\frac{\partial K}{\partial p_i}\cdot dp_i + \sum_{i=1}^{N}\left(\frac{\partial L}{\partial q_i}\right)\cdot dq_i - \sum_{i=1}^{N} v_i \cdot dp_i = 0. \tag{11.9}$$

Combing the expressions with the same differentials we get:

$$\sum_{i=1}^{N}\left(\frac{\partial K}{\partial q_i}+\frac{\partial L}{\partial q_i}\right)\cdot dq_i + \sum_{i=1}^{N}\left(\frac{\partial K}{\partial p_i}-v_i\right)\cdot dp_i = 0. \tag{11.10}$$

Before we use (11.10), let us recall that it was obtained with the condition that the Legendre Transformations were satisfied. If we look at the Legendre Transformations written in the form (9.14), we have:

$$\varphi_s(q,p) = 0, \qquad s = M+1,\ldots,N. \tag{11.11}$$

(We skipped the other equations from (9.14), since here we are interested in looking only at relations between dq_i and dp_i that appear in (11.10)).

Calculating the differentials of φ_s, $s = M+1,...,N$, from (11.11) we get:

$$d\varphi_s = \sum_{i=1}^{N} \frac{\partial \varphi_s}{\partial q_i} \cdot dq_i + \sum_{i=1}^{N} \frac{\partial \varphi_s}{\partial p_i} \cdot dp_i, \quad s = M+1,...,N. \tag{11.12}$$

However, we restrict ourselves to the points that satisfy the Legendre Transformations (11.11). For these points, we have $\varphi_s(q,p) = 0$, so the infinitesimal change of it are:

$$d\varphi_s = 0, \quad s = M+1,...,N.$$

Therefore, from (11.12) we get:

$$\sum_{i=1}^{N} \frac{\partial \varphi_s}{\partial q_i} \cdot dq_i + \sum_{i=1}^{N} \frac{\partial \varphi_s}{\partial p_i} \cdot dp_i = 0, \quad s = M+1,...,N. \tag{11.13}$$

The formulas (11.13) are the conditions that infinitesimal changes dq_i and dp_i must satisfy. In other words, starting with a point satisfying the Legendre Transformations, we move, in the infinitesimal sense, to another point that is still satisfying the Legendre Transformations.

We will not go into all details here, but for (11.13) to be a sufficient condition for the correct infinitesimal changes dq_i and dp_i, the form of (11.11) must be such that the partial derivatives do not become zero automatically on the constraints. For example, $\varphi_1 = q_1^2 = 0$, is not in the correct form, while $\phi_1 = q_1 = 0$ is, despite the fact that both give really the same condition that $q_1 = 0$.

This is because $d\varphi_1 = \frac{\partial (q_1^2)}{\partial q_1} dq_1 = 2q_1 dq_1 = 2 \cdot 0 \, dq_1 = 0$ is not creating any conditions for dq_1,

while $d\phi_1 = \frac{\partial q_1}{\partial q_1} dq_1 = 1 dq_1 = 0$ gives the correct condition, $dq_1 = 0$. Without going into much details, let us just say that a correct form for a constraints can be achieved by solving the original constraint for one of the variables, and then obtaining zero on one side by addition or subtraction to both sides of the constraint.

Now, let us fix a point (q, p, v) satisfying the Legendre Transformations, and at this point, introduce the following:

1) V is an n-dimensional vector space, with $n = 2N$

2) A set of vectors $\{\vec{w}_{M+1}, \vec{w}_{M+2}, ..., \vec{w}_N\}$ in V, given as:

$$\vec{w}_s = \left(\frac{\partial \varphi_s}{\partial q_1}, \frac{\partial \varphi_s}{\partial q_N}, ..., \frac{\partial \varphi_s}{\partial q_N}, \frac{\partial \varphi_s}{\partial p_1}, \frac{\partial \varphi_s}{\partial p_1}, ..., \frac{\partial \varphi_s}{\partial p_N} \right), \qquad s = M+1, ..., N. \tag{11.14}$$

3) U is the set of all vectors $\vec{u}$ from V given as:

$$\vec{u} = A(dq_1, dq_2, ..., dq_N, dp_1, dp_2, ..., dp_N), \tag{11.15}$$

such that $(dq_1, dq_2, ..., dq_N, dp_1, dp_2, ..., dp_N) \cdot \vec{w}_s = 0, \qquad s = M+1, ..., N.$ (11.16)

The symbol A in (11.15) is not a real number and placing it in front of the infinitesimal vector $(dq_1, dq_2, ..., dq_N, dp_1, dp_2, ..., dp_N)$ does not mean multiplication. Real number multiplied by the infinitesimal vector $(dq_1, dq_2, ..., dq_N, dp_1, dp_2, ..., dp_N)$, would produce another infinitesimal vector, and this is not what we want here. Rather, the entire symbol $A(dq_1, dq_2, ..., dq_N, dp_1, dp_2, ..., dp_N)$ denotes an arbitrary non-infinitesimal vector such that when a dot multiplied by any vector $\vec{w}_s$ gives zero. By writing $A(dq_1, dq_2, ..., dq_N, dp_1, dp_2, ..., dp_N)$, we symbolically express the fact that such vectors will have coordinates that have the same proportions between themselves as the infinitesimal vectors satisfying (11.16).

Since the condition (11.16), which is identical to (11.13), is giving us all infinitesimal vectors giving zeros when dot multiplied by each vector $\vec{w}_s$, the vectors (11.15) give us all non-infinitesimal vectors that give zero when multiplied by all $\vec{w}_s$. So, the set U is a set of all vectors

$\vec{u}$ that satisfy:

$$\vec{u} \cdot \vec{w}_{M+1} = 0,$$

$$\vec{u} \cdot \vec{w}_{M+2} = 0,$$

$$\vdots$$

$$\vec{u} \cdot \vec{w}_N = 0.$$

4) Define the vector $\vec{x}$ as:

$$\vec{x} = \left(\frac{\partial K}{\partial q_1} + \frac{\partial L}{\partial q_1}, \frac{\partial K}{\partial q_2} + \frac{\partial L}{\partial q_2}, \ldots, \frac{\partial K}{\partial q_N} + \frac{\partial L}{\partial q_N}, \frac{\partial K}{\partial p_1} - v_1, \frac{\partial K}{\partial p_2} - v_2, \ldots, \frac{\partial K}{\partial p_N} - v_N \right). \qquad (11.17)$$

Notice that, because of (11.10), we have:

$$\vec{x} \cdot (dq_1, dq_2, \ldots, dq_N, dp_1, dp_2, \ldots, dp_N) = 0 \qquad (11.18)$$

So, we also have:

$$\vec{x} \cdot \left[A(dq_1, dq_2, \ldots, dq_N, dp_1, dp_2, \ldots, dp_N) \right] = 0. \qquad (11.19)$$

Since $A(dq_1, dq_2, \ldots, dq_N, dp_1, dp_2, \ldots, dp_N)$ gives all vectors in the set U, we have $\vec{x} \cdot \vec{u} = 0$, for all vectors in U.

Notice then the conditions 1) – 4) above are identical to the assumptions 1) – 4) of the main theorem given in Chapter X. Therefore, we get the conclusion of the theorem, that $\vec{x}$ is a linear combination of the vectors $\{\vec{w}_{M+1}, \vec{w}_{M+2}, \ldots, \vec{w}_N\}$. In other words, we have:

$$\vec{x} = \sum_{s=M+1}^{N} (-u_s) \vec{w}_s, \qquad (11.20)$$

where u_s are some real numbers (We use minuses in front of them to get a simpler final answer.)

Using (11.17) and (11.14) in (11.20) we get:

$$\left(\frac{\partial K}{\partial q_1}+\frac{\partial L}{\partial q_1},\frac{\partial K}{\partial q_2}+\frac{\partial L}{\partial q_2},\ldots,\frac{\partial K}{\partial q_N}+\frac{\partial L}{\partial q_N},\frac{\partial K}{\partial p_1}-v_1,\frac{\partial K}{\partial p_2}-v_2,\ldots,\frac{\partial K}{\partial p_N}-v_N\right)=$$

$$=\sum_{s=M+1}^{N}-u_s\left(\frac{\partial \varphi_s}{\partial q_1},\frac{\partial \varphi_s}{\partial q_N},\ldots,\frac{\partial \varphi_s}{\partial q_N},\frac{\partial \varphi_s}{\partial p_1},\frac{\partial \varphi_s}{\partial p_1},\ldots,\frac{\partial \varphi_s}{\partial p_N}\right). \tag{11.21}$$

Multiplying out the bottom line and moving the additions to each component inside of the vectors, we get:

$$\left(\frac{\partial K}{\partial q_1}+\frac{\partial L}{\partial q_1},\frac{\partial K}{\partial q_2}+\frac{\partial L}{\partial q_2},\ldots,\frac{\partial K}{\partial q_N}+\frac{\partial L}{\partial q_N},\frac{\partial K}{\partial p_1}-v_1,\frac{\partial K}{\partial p_2}-v_2,\ldots,\frac{\partial K}{\partial p_N}-v_N\right)=$$

$$=\left(\sum_{s=M+1}^{N}-u_s\frac{\partial \varphi_s}{\partial q_1},\sum_{s=M+1}^{N}-u_s\frac{\partial \varphi_s}{\partial q_2},\ldots,\sum_{s=M+1}^{N}-u_s\frac{\partial \varphi_s}{\partial q_N},\sum_{s=M+1}^{N}-u_s\frac{\partial \varphi_s}{\partial p_1},\sum_{s=M+1}^{N}-u_s\frac{\partial \varphi_s}{\partial p_2},\ldots,\sum_{s=M+1}^{N}-u_s\frac{\partial \varphi_s}{\partial p_N}\right).$$

Comparing the vectors component by component we get:

$$\frac{\partial K}{\partial q_1}+\frac{\partial L}{\partial q_1}=\sum_{s=M+1}^{N}-u_s\frac{\partial \varphi_s}{\partial q_1},$$

$$\frac{\partial K}{\partial q_2}+\frac{\partial L}{\partial q_2}=\sum_{s=M+1}^{N}-u_s\frac{\partial \varphi_s}{\partial q_2},$$

$$\vdots$$

$$\frac{\partial K}{\partial q_N}+\frac{\partial L}{\partial q_N}=\sum_{s=M+1}^{N}-u_s\frac{\partial \varphi_s}{\partial q_N},$$

$$\frac{\partial K}{\partial p_1}-v_1=\sum_{s=M+1}^{N}-u_s\frac{\partial \varphi_s}{\partial p_1},$$

$$\frac{\partial K}{\partial p_2}-v_2=\sum_{s=M+1}^{N}-u_s\frac{\partial \varphi_s}{\partial p_2},$$

$$\vdots$$

$$\frac{\partial K}{\partial p_N}-v_N=\sum_{s=M+1}^{N}-u_s\frac{\partial \varphi_s}{\partial p_N}. \tag{11.22}$$

When we solve the equations (11.22) for either $\frac{\partial L}{\partial q_i}$ or v_1, $i = 1,...,N$, we get:

$$\frac{\partial L}{\partial q_1} = -\frac{\partial K}{\partial q_1} - \sum_{s=M+1}^{N} u_s \frac{\partial \varphi_s}{\partial q_1},$$

$$\frac{\partial L}{\partial q_2} = -\frac{\partial K}{\partial q_2} - \sum_{s=M+1}^{N} u_s \frac{\partial \varphi_s}{\partial q_2},$$

$$\vdots$$

$$\frac{\partial L}{\partial q_N} = -\frac{\partial K}{\partial q_N} - \sum_{s=M+1}^{N} u_s \frac{\partial \varphi_s}{\partial q_N},$$

$$v_1 = \frac{\partial K}{\partial p_1} + \sum_{s=M+1}^{N} u_s \frac{\partial \varphi_s}{\partial p_1}, \tag{11.23}$$

$$v_2 = \frac{\partial K}{\partial p_2} + \sum_{s=M+1}^{N} u_s \frac{\partial \varphi_s}{\partial p_2},$$

$$\vdots$$

$$v_N = \frac{\partial K}{\partial p_N} + \sum_{s=M+1}^{N} u_s \frac{\partial \varphi_s}{\partial p_N}.$$

Formulas (11.23) were obtained for one specific (q,p,v). However, we can repeat it for any (q,p,v), getting:

$$\frac{\partial L}{\partial q_1} = -\frac{\partial K(q,p)}{\partial q_1} - \sum_{s=M+1}^{N} u_s(q,p,v) \frac{\partial \varphi_s(q,p)}{\partial q_1},$$

$$\frac{\partial L}{\partial q_2} = -\frac{\partial K(q,p)}{\partial q_2} - \sum_{s=M+1}^{N} u_s(q,p,v) \frac{\partial \varphi_s(q,p)}{\partial q_2},$$

$$\vdots$$

$$\frac{\partial L}{\partial q_N} = -\frac{\partial K(q,p)}{\partial q_N} - \sum_{s=M+1}^{N} u_s(q,p,v) \frac{\partial \varphi_s(q,p)}{\partial q_N},$$

$$v_1 = \frac{\partial K(q,p)}{\partial p_1} + \sum_{s=M+1}^{N} u_s(q,p,v) \frac{\partial \varphi_s(q,p)}{\partial p_1},$$

(continued on the next page)

$$v_2 = \frac{\partial K(q,p)}{\partial p_2} + \sum_{s=M+1}^{N} u_s(q,p,v) \frac{\partial \varphi_s(q,p)}{\partial p_2},$$
$$\vdots$$
$$v_N = \frac{\partial K(q,p)}{\partial p_N} + \sum_{s=M+1}^{N} u_s(q,p,v) \frac{\partial \varphi_s(q,p)}{\partial p_N}.$$

(11.24)

Notice that the variables (q,p,v) show up in u_s, but not in K and φ_s. This is because we established earlier that both K and φ_s are functions of (q,p) only.

Finally, we observe that the functions u_s will be used to modify the pre-Hamiltonian K so that the modified pre-Hamiltonian will produce the Hamilton's equations of motion that will match the original Euler-Lagrange equations of motion. However, as shown in Chapter V, the original Euler – Lagrange equations of motion always produce the relations (5.13) that always express velocities in terms of positions in momenta. (This is done not by the Legendre Transformations, but by secondary constraints produced by the equations of motion.) Then, since we intend to use the formulas (11.24) only for the solutions of the equations of motion, we can use (5.14), redefining u_s as:

$$u_s(q,p) = u_s(q,p,v(q,p)),$$

(11.25)

where $v(q,p)$ is taken from (5.14). So, we get:

$$\frac{\partial L}{\partial q_1} = -\frac{\partial K(q,p)}{\partial q_1} - \sum_{s=M+1}^{N} u_s(q,p) \frac{\partial \varphi_s(q,p)}{\partial q_1},$$
$$\frac{\partial L}{\partial q_2} = -\frac{\partial K(q,p)}{\partial q_2} - \sum_{s=M+1}^{N} u_s(q,p) \frac{\partial \varphi_s(q,p)}{\partial q_2},$$
$$\vdots$$
$$\frac{\partial L}{\partial q_N} = -\frac{\partial K(q,p)}{\partial q_N} - \sum_{s=M+1}^{N} u_s(q,p) \frac{\partial \varphi_s(q,p)}{\partial q_N},$$

(continued on the next page)

$$v_1 = \frac{\partial K(q,p)}{\partial p_1} + \sum_{s=M+1}^{N} u_s(q,p) \frac{\partial \varphi_s(q,p)}{\partial p_1},$$

$$v_2 = \frac{\partial K(q,p)}{\partial p_2} + \sum_{s=M+1}^{N} u_s(q,p) \frac{\partial \varphi_s(q,p)}{\partial p_2}, \qquad (11.26)$$

$$\vdots$$

$$v_N = \frac{\partial K(q,p)}{\partial p_N} + \sum_{s=M+1}^{N} u_s(q,p) \frac{\partial \varphi_s(q,p)}{\partial p_N}.$$

The equations (11.26) may look strange since the variables on the left sides are (q,v) and the variables on the right sides are (q,p,v). However, there is nothing wrong with them, since they are supposed to hold only for the points satisfying the Legendre Transformations, and the Legendre Transformations itself have momenta on one side and positions and velocities on the other. So, the left side can still be equal to the right side.

Also, notice that the theorem we used to get (11.26) does not say it will work for any arbitrary u_s. It only says that such u_s exists. As a consequence, the formula (11.26) only says that there exists such $u_s(q,p)$ that (11.26) holds.

2. The Hamilton's Equations for a Constrained Lagrangian

Combining the equations (11.26) with the Euler-Lagrange equations (2.2) from Chapter II, we get:

$$\frac{d}{dt}\left[\frac{\partial L}{\partial v_1}\right] = -\frac{\partial K(q,p)}{\partial q_1} - \sum_{s=M+1}^{N} u_s(q,p) \frac{\partial \varphi_s(q,p)}{\partial q_1},$$

$$\frac{d}{dt}\left[\frac{\partial L}{\partial v_2}\right] = -\frac{\partial K(q,p)}{\partial q_2} - \sum_{s=M+1}^{N} u_s(q,p) \frac{\partial \varphi_s(q,p)}{\partial q_2},$$

$$\vdots$$

$$\frac{d}{dt}\left[\frac{\partial L}{\partial v_N}\right] = -\frac{\partial K(q,p)}{\partial q_N} - \sum_{s=M+1}^{N} u_s(q,p) \frac{\partial \varphi_s(q,p)}{\partial q_N},$$

(continued on the next page)

$$\frac{dq_1}{dt} = \frac{\partial K(q,p)}{\partial p_1} + \sum_{s=M+1}^{N} u_s(q,p)\frac{\partial \varphi_s(q,p)}{\partial p_1},$$

$$\frac{dq_2}{dt} = \frac{\partial K(q,p)}{\partial p_2} + \sum_{s=M+1}^{N} u_s(q,p)\frac{\partial \varphi_s(q,p)}{\partial p_2}, \qquad (11.27)$$

$$\vdots$$

$$\frac{dq_N}{dt} = \frac{\partial K(q,p)}{\partial p_N} + \sum_{s=M+1}^{N} u_s(q,p)\frac{\partial \varphi_s(q,p)}{\partial p_N}.$$

Using the Legendre Transformations $p_i = \frac{\partial L}{\partial v_i}$, we get:

$$\frac{dp_1}{dt} = -\frac{\partial K(q,p)}{\partial q_1} - \sum_{s=M+1}^{N} u_s(q,p)\frac{\partial \varphi_s(q,p)}{\partial q_1},$$

$$\frac{dp_2}{dt} = -\frac{\partial K(q,p)}{\partial q_2} - \sum_{s=M+1}^{N} u_s(q,p)\frac{\partial \varphi_s(q,p)}{\partial q_2},$$

$$\frac{dp_N}{dt} = -\frac{\partial K(q,p)}{\partial q_N} - \sum_{s=M+1}^{N} u_s(q,p)\frac{\partial \varphi_s(q,p)}{\partial q_N},$$

$$\frac{dq_1}{dt} = \frac{\partial K(q,p)}{\partial p_1} + \sum_{s=M+1}^{N} u_s(q,p)\frac{\partial \varphi_s(q,p)}{\partial p_1}, \qquad (11.28)$$

$$\frac{dq_2}{dt} = \frac{\partial K(q,p)}{\partial p_2} + \sum_{s=M+1}^{N} u_s(q,p)\frac{\partial \varphi_s(q,p)}{\partial p_2},$$

$$\vdots$$

$$\frac{dq_N}{dt} = \frac{\partial K(q,p)}{\partial p_N} + \sum_{s=M+1}^{N} u_s(q,p)\frac{\partial \varphi_s(q,p)}{\partial p_N}.$$

Now, for the points that satisfy the Legendre Transforms, we have:

$$\sum_{s=M+1}^{N} \frac{\partial u_s(q,p)}{\partial q_i} \cdot \varphi_s(q,p) = 0, \qquad i = 1,\ldots N. \qquad (11.29)$$

since $\varphi_s(q,p) = 0$ for the points that satisfy the Legendre Transforms.

For the same reason, we have:

$$\sum_{s=M+1}^{N}\frac{\partial u_s(q,p)}{\partial p_i}\cdot\varphi_s(q,p)=0, \qquad i=1,\ldots N. \tag{11.30}$$

Including (11.29) and (11.30) in the proper equations in (11.28) gives:

$$\begin{aligned}
\frac{dp_1}{dt} &= -\frac{\partial K(q,p)}{\partial q_1} - \sum_{s=M+1}^{N} u_s(q,p)\frac{\partial \varphi_s(q,p)}{\partial q_1} - \sum_{s=M+1}^{N}\frac{\partial u_s(q,p)}{\partial q_1}\cdot\varphi_s(q,p), \\
\frac{dp_2}{dt} &= -\frac{\partial K(q,p)}{\partial q_2} - \sum_{s=M+1}^{N} u_s(q,p)\frac{\partial \varphi_s(q,p)}{\partial q_2} - \sum_{s=M+1}^{N}\frac{\partial u_s(q,p)}{\partial q_2}\cdot\varphi_s(q,p), \\
&\vdots \\
\frac{dp_N}{dt} &= -\frac{\partial K(q,p)}{\partial q_N} - \sum_{s=M+1}^{N} u_s(q,p)\frac{\partial \varphi_s(q,p)}{\partial q_N} - \sum_{s=M+1}^{N}\frac{\partial u_s(q,p)}{\partial q_N}\cdot\varphi_s(q,p), \\
\frac{dq_1}{dt} &= \frac{\partial K(q,p)}{\partial p_1} + \sum_{s=M+1}^{N} u_s(q,p)\frac{\partial \varphi_s(q,p)}{\partial p_1} + \sum_{s=M+1}^{N}\frac{\partial u_s(q,p)}{\partial p_1}\cdot\varphi_s(q,p), \\
\frac{dq_2}{dt} &= \frac{\partial K(q,p)}{\partial p_2} + \sum_{s=M+1}^{N} u_s(q,p)\frac{\partial \varphi_s(q,p)}{\partial p_2} + \sum_{s=M+1}^{N}\frac{\partial u_s(q,p)}{\partial p_2}\cdot\varphi_s(q,p), \\
&\vdots \\
\frac{dq_N}{dt} &= \frac{\partial K(q,p)}{\partial p_N} + \sum_{s=M+1}^{N} u_s(q,p)\frac{\partial \varphi_s(q,p)}{\partial p_N} + \sum_{s=M+1}^{N}\frac{\partial u_s(q,p)}{\partial p_N}\cdot\varphi_s(q,p).
\end{aligned} \tag{11.31}$$

Combining the summations and using the product rule, we get:

$$\begin{aligned}
\frac{dp_1}{dt} &= -\frac{\partial K(q,p)}{\partial q_1} - \sum_{s=M+1}^{N}\frac{\partial}{\partial q_1}\left[u_s(q,p)\varphi_s(q,p)\right], \\
\frac{dp_2}{dt} &= -\frac{\partial K(q,p)}{\partial q_2} - \sum_{s=M+1}^{N}\frac{\partial}{\partial q_2}\left[u_s(q,p)\varphi_s(q,p)\right], \\
&\vdots \\
\frac{dp_N}{dt} &= -\frac{\partial K(q,p)}{\partial q_N} - \sum_{s=M+1}^{N}\frac{\partial}{\partial q_N}\left[u_s(q,p)\varphi_s(q,p)\right], \\
\frac{dq_1}{dt} &= \frac{\partial K(q,p)}{\partial p_1} + \sum_{s=M+1}^{N}\frac{\partial}{\partial p_1}\left[u_s(q,p)\varphi_s(q,p)\right],
\end{aligned}$$

(continued on the next page)

$$\frac{dq_2}{dt} = \frac{\partial K(q,p)}{\partial p_2} + \sum_{s=M+1}^{N} \frac{\partial}{\partial p_2}\left[u_s(q,p)\varphi_s(q,p)\right],$$
$$\vdots$$
$$\frac{dq_N}{dt} = \frac{\partial K(q,p)}{\partial p_N} + \sum_{s=M+1}^{N} \frac{\partial}{\partial p_N}\left[u_s(q,p)\varphi_s(q,p)\right].$$
(11.32)

Now we can combine the partial derivatives in each line into just one derivative, getting:

$$\frac{dp_1}{dt} = -\frac{\partial}{\partial q_1}\left[K(q,p) + \sum_{s=M+1}^{N} u_s(q,p)\varphi_s(q,p)\right],$$
$$\frac{dp_2}{dt} = -\frac{\partial}{\partial q_2}\left[K(q,p) + \sum_{s=M+1}^{N} u_s(q,p)\varphi_s(q,p)\right],$$
$$\vdots$$
$$\frac{dp_N}{dt} = -\frac{\partial}{\partial q_N}\left[K(q,p) + \sum_{s=M+1}^{N} u_s(q,p)\varphi_s(q,p)\right],$$
$$\frac{dq_1}{dt} = \frac{\partial}{\partial p_1}\left[K(q,p) + \sum_{s=M+1}^{N} u_s(q,p)\varphi_s(q,p)\right],$$
$$\frac{dq_2}{dt} = \frac{\partial}{\partial p_2}\left[K(q,p) + \sum_{s=M+1}^{N} u_s(q,p)\varphi_s(q,p)\right],$$
$$\vdots$$
$$\frac{dq_N}{dt} = \frac{\partial}{\partial p_N}\left[K(q,p) + \sum_{s=M+1}^{N} u_s(q,p)\varphi_s(q,p)\right].$$
(11.33)

Again, let us repeat that the formulas (11.33) will not work for all $u_s(q,p)$. Instead, the formulas (11.33) only say that there exist such $u_s(q,p)$ that the equations (11.33) are equivalent to the original Euler-Lagrange equations (2.2) from Chapter II.

In practice, when we want to calculate the $u_s(q,p)$ that will give the equations (11.33) equivalent to the Euler-Lagrange equations, we use the first two lines of the Hamilton's equations (5.14) from Chapter V and combine them with equations (11.28). We obtain:

$$F_i(q,p) = -\frac{\partial K(q,p)}{\partial q_i} - \sum_{s=M+1}^{N} u_s(q,p)\frac{\partial \varphi_s(q,p)}{\partial q_i},$$

$$v_i(q,p) = \frac{\partial K(q,p)}{\partial p_i} + \sum_{s=M+1}^{N} u_s(q,p)\frac{\partial \varphi_s(q,p)}{\partial p_i}, \qquad i = 1,\ldots,N. \tag{11.34}$$

The fact that earlier we proved the existence of $u_s(q,p)$ that gives the Euler-Lagrange equations back, means that the equations (11.34) can in principle be solved for $u_s(q,p)$. However, in practice it can be difficult or impossible.

The solution for $u_s(q,p)$ is, in general, not unique.

Once we get the correct $u_s(q,p)$, then we call (11.34) the Hamilton's equations for our system.

3. The Definition of the Hamiltonian

After we get such $u_s(q,p)$, $s = M+1,\ldots,N$, so that we get the correct Hamilton's equations (11.34), we define the Hamiltonian as:

$$H(q,p) = K(q,p) + \sum_{s=M+1}^{N} u_s(q,p)\varphi_s(q,p). \tag{11.35}$$

Notice that in the formula of the Hamiltonian (11.35), we only use the constraints $\varphi_s(q,p)$ obtained directly from the Legendre Transformations. These constraints were earlier called the primary constraints.

Using the Hamiltonian (11.35) we can see that the Hamilton's equations (11.33) become: (we use the dot above a variable to denote the time derivative and we include the primary constraints):

$$\dot{p}_i = -\frac{\partial H}{\partial q_i},$$

(continued on the next page)

$$\dot{q}_i = \frac{\partial H}{\partial p_i}, \qquad i = 1,\ldots,N,$$
$$\varphi_s(q,p) = 0, \qquad s = M+1,\ldots,N. \tag{11.36}$$

In (11.36) we used the dot above a variable to denote the time derivative, and we included the primary constraints.

The equations (11.36) will also be called the Hamilton's equations.

From the equations (3.14), by looking at the subsequent time derivatives of the constraints, we will get the same constraints as in (5.14), in Chapter V. Including all the constraints we can write the equations (11.36) as:

$$\dot{p}_i = -\frac{\partial H}{\partial q_i},$$
$$\dot{q}_i = \frac{\partial H}{\partial p_i}, \qquad i = 1,\ldots,N,$$
$$\varphi_s(q,p) = 0, \qquad s = 1,\ldots,S. \tag{11.37}$$

The (11.37) are also called the Hamilton's equations. Now they include all constraints of the system.

Since the first two lines in (11.37) have the form identical to the equations (3.14) from the Chapter III, then when we introduce the usual Poisson Brackets as in (3.19):

$$\{F,G\} = \sum_{i=1}^{N}\left(\frac{\partial F}{\partial q_i}\cdot\frac{\partial G}{\partial p_i} - \frac{\partial F}{\partial p_i}\cdot\frac{\partial G}{\partial q_i}\right), \tag{11.38}$$

we get:

$$\dot{p}_i = \{p_i, H\},$$
$$\dot{q}_i = \{q_i, H\}, \qquad i = 1,\ldots,N,$$
$$\varphi_s(q,p) = 0, \qquad s = 1,\ldots,S. \tag{11.39}$$

The only difference, compared to the hyperregular case, is the existence of the constraints in (11.39). This also means that the equations of motion (11.39) are holding only for the points satisfying the constraints.

The equations (11.39) are also called the Hamilton's equations.

The same way as in the hyperregular case, for any function $F = F(q,p)$, we have:

$$\begin{aligned} \dot{F} &= \{F,H\}, \\ \varphi_s(q,p) &= 0, \quad s = 1,...,S. \end{aligned} \qquad (11.40)$$

4. The relation between the pre-Hamiltonian and the Hamiltonian

Comparing the definition (11.35) of the Hamiltonian,

$$H(q,p) = K(q,p) + \sum_{s=M+1}^{N} u_s(q,p)\varphi_s(q,p),$$

with the results of Chapter VI, we conclude that the Hamiltonian can be obtained from the pre-Hamiltonian by a generalized substitution that uses the primary constraints. One conclusion that follows is that the Hamiltonian and the pre-Hamiltonian are identical on the points that satisfy the primary constraints. The second conclusion is that, instead of trying to find the functions $u_s(q,p)$, we may try some substitutions using the primary constraints. In practice, we can often do it by guessing, and checking, by comparing the Hamilton's equations (11.39) obtained from the guessed Hamiltonian, with the Hamilton's equations (5.14) obtained directly from the Euler-Lagrange equations.

Finally, we need to stress that not every substitution using constraints in the Hamiltonian will work in the sense of giving us the correct equations of motion. An example of this was presented in Chapter IV.

5. Including the Secondary Constraints in the Hamiltonian

We may try to include the secondary constraints in the Hamiltonian as well. They are not needed, but they may be convenient to use. Notice that, when showing that the pre-Hamiltonian K is a function of (q, p) only, we used the method of reconstruction of a function from its differential. However, in practical calculations, the removal of velocities from K may be much simpler if we use the formulas for expressing velocities by positions and momenta from the formulas (5.14) from Chapter V. These formulas must be obtained anyway to get the correct Hamiltonian. Using them as substitutions may be the simplest way to remove the velocities from the pre-Hamiltonian. Let us stress, though, that removing the velocities from the pre-Hamiltonian will not, in general, give us the correct Hamiltonian right away.

Using the above will give the same Hamiltonian on all points satisfying all constraints, and then, as we know from Chapter VI, the Hamiltonian (11.35) can be obtained from it, by adding linear combinations of all constraints. This way, we get the Hamiltonian (11.35) as

$$H(q,p) = K(q,p) + \sum_{s=1}^{S} u_s(q,p)\varphi_s(q,p). \tag{11.41}$$

This formula may look identical to (11.35), but now we use all constraints in it, not just the primary constraints. Since we have possible freedom in choosing $u_s(q,p)$, this formula may give a different Hamiltonian than the formula (11.35), but since we still require that it produces the same equations of motion (the equations of motion must hold on all constraints only), we accept it as a correct Hamiltonian, possibly different from the one obtained using the primary constraints only.

As before, the formula (11.41) is not saying that any $u_s(q,p)$ in it will work in the sense of producing the correct equations of motion. It rather says that there exists such $u_s(q,p)$, not necessarily unique, that the Hamiltonian (11.41) will produce the correct equations of motion.

Let us stress that in all our considerations we always expect that the Hamilton's equations, which we obtain at the end of the process, are equivalent to the original Euler-Lagrange equations obtained directly from the Lagrangian.

CHAPTER XII

A SUMMARY – GETTING THE HAMILTONIAN FORMALISM FROM A REGULAR CONSTRAINED LAGRANGIAN

In the previous chapters, we were making efforts to prove every statement with as much details as reasonably possible. So, it is likely that the general picture got somewhat lost. Thus, in this chapter we want to give an overview of what was done, and how it can be applied. We will start with a Lagrangian and describe a procedure of obtaining the full Lagrange and Hamiltonian structure, skipping the many details that were given in the previous chapters.

1. Getting the Hamiltonian Formalism from a Regular Constrained Lagrangian

As before, let us assume that we have an N - dimensional mechanical system, described by the position-velocity variables $(q,v) = (q_1, q_2, ..., q_N, v_1, v_2, ..., v_N)$ and a constrained Lagrangian:

$$L = L(q,v). \tag{12.1}$$

The Lagrangian (12.1) produces so called Euler-Lagrange differential equations of motion:

$$\frac{d}{dt}\left[\frac{\partial L}{\partial v_i}\right] = \frac{\partial L}{\partial q_i},$$
$$\frac{dq_i}{dt} = v_i, \qquad i = 1, ..., N. \tag{12.2}$$

Then, we use the Legendre Transformations to introduce the momenta p_i:

$$p_i = \frac{\partial L}{\partial v_i}, \qquad i = 1, ..., N. \tag{12.3}$$

The Legendre Transformations (12.3) can, in principle, be solved for as many velocities as possible, giving:

$$v_k = v_k(q, p, v_{M+1}, v_{M+2}, ..., v_N), \quad k = 1, ..., M < N,$$
$$\varphi_s(q, p) = 0, \quad s = M+1, ..., N. \tag{12.4}$$

In the above, without loss of generality, we assumed that we solved for the first M velocities. We also obtained exactly $N - M$ independent relations between positions and momenta. These relations were called the primary constraints.

Also, to avoid showing the partial derivatives symbols, we define:

$$F_i = F_i(q, v) = \left.\frac{\partial L}{\partial q_i}\right|_{(q,v)}, \quad i = 1, ..., N. \tag{12.5}$$

Then, using (12.5), with a dot above a variable denoting the time derivative of the variable, using the definitions of momenta (12.3) and including (12.4) the complete equations of motion of the system can be written as:

$$\dot{p}_i = F_i(q, v), \quad i = 1, ..., N, \tag{12.6}$$

$$\dot{q}_i = v_i, \quad i = 1, ..., N, \tag{12.7}$$

$$v_k = v_k(q, p, v_{M+1}, v_{M+2}, ..., v_N), \quad k = 1, ..., M < N, \tag{12.8}$$

$$\varphi_s(q, p) = 0, \quad s = M+1, ..., N. \tag{12.9}$$

Then, calculating subsequent time derivatives of the constraints in (12.9), we may either obtain identities that bring no new information or equations that give us more velocities as in (12.8), or more constraints as in (12.9). We can also get a contradiction, which would tell us that our Lagrangian should be discarded.

Then, we take the time derivatives of all newly obtained constraints and repeat the same process, until no new constraints appear. The process is finite, since the number of independent constraints cannot be larger than $2N$.

When the process is finished, and no contradictions appear, we get:

$$\begin{aligned}&\dot{p}_i = F_i(q,p), \\ &\dot{q}_i = v_i(q,p), \\ &v_i = v_i(q,p), \qquad\qquad i = 1,\ldots,N, \\ &\varphi_s(q,p) = 0, \qquad\qquad s = 1,\ldots,S.\end{aligned} \qquad (12.10)$$

We assume that all velocities will be expressed by (q,p) at the end of this process. This assumption is a separate assumption about the Lagrangian. If that assumption is not satisfied, then the time evolution of some velocities, and consequently positions and momenta, would not be given by their initial values, and we consider this situation to be outside the scope of this book. If a Lagrangian produces equations in which this assumption is satisfied, we call the Lagrangian regular. In this book, we only consider regular Lagrangians.

The constraints in the last line of (12.10), represent the complete set of all constraints in the system, both primary and secondary. We will assume that these constraints are independent, meaning that removing any one of them would change the points (q,p) which satisfy the constraints in (12.10). The details of making the constraints independent were given in Chapter V, when obtaining the formulas (5.12).

In (12.10), we can drop the third line since all the information about it is contained in line two. The system we get is:

$$\begin{aligned}&\dot{p}_i = F_i(q,p), \\ &\dot{q}_i = v_i(q,p), \qquad\qquad i = 1,\ldots,N, \\ &\varphi_s(q,p) = 0, \qquad\qquad s = 1,\ldots,S.\end{aligned} \qquad (12.11)$$

We define a pre-Hamiltonian K as:

$$K = K(q,v,p) = \sum_{i=1}^{N} p_i \cdot v_i - L(q,v). \qquad (12.12)$$

We can use the formulas from (12.10) to eliminate velocities from K in (12.12). So, we get:

$$K(q,v,p) = K(q,p). \tag{12.13}$$

Then we define the Hamiltonian, as:

$$H(q,p) = K(q,p) + \sum_{s=1}^{S} u_s(q,p)\varphi_s(q,p), \tag{12.14}$$

where we may use the primary constraints only, or possibly also other constraints of the system, whatever is more convenient. The functions $u_s(q,p)$ in (12.14) are chosen in such a way that:

$$\begin{aligned}F_i(q,p) &= -\frac{\partial H}{\partial q_i}, \\ v_i(q,p) &= \frac{\partial H}{\partial p_i}, \qquad i = 1,\ldots,N.\end{aligned} \tag{12.15}$$

We have shown that such functions $u_s(q,p)$ always exist, even if only the primary constraints are used in (12.14). Obviously, they also exist when using all constraints, since we could multiply non-primary constraints by zeros. Still, other choices for $u_s(q,p)$ may exist and be more convenient. In general, the functions $u_s(q,p)$ are not unique.

Notice that we need to get the equations (12.11) first, before we are able to get the Hamiltonian that satisfies (12.15), since we need to know the left sides in (12.15) to get the correct $u_s(q,p)$ in (12.14).

Once we have the Hamiltonian that satisfies (12.15), we compare it with the equations (12.11), and we get:

$$\dot{p}_i = -\frac{\partial H}{\partial q_i},$$
$$\dot{q}_i = \frac{\partial H}{\partial p_i}, \qquad i = 1,\dots,N, \qquad (12.16)$$
$$\varphi_s(q,p) = 0, \qquad s = 1,\dots,S.$$

The equations (12.16) are called the Hamilton's equations.

We introduce the Poisson Bracket, defined as:

$$\{F,G\} = \sum_{i=1}^{N} \left(\frac{\partial F}{\partial q_i} \cdot \frac{\partial G}{\partial p_i} - \frac{\partial F}{\partial p_i} \cdot \frac{\partial G}{\partial q_i} \right). \qquad (12.17)$$

Then (12.16) can be written as:

$$\dot{p}_i = \{p_i, H\},$$
$$\dot{q}_i = \{q_i, H\}, \qquad i = 1,\dots,N, \qquad (12.18)$$
$$\varphi_s(q,p) = 0, \qquad s = 1,\dots,S.$$

Then for any function $F = F(q,p)$, we have:

$$\dot{F} = \{F, H\},$$
$$\varphi_s(q,p) = 0, \qquad s = 1,\dots,S. \qquad (12.19)$$

Let us repeat here that the constraints in (12.10), (12.11), (12.16), (12.18), and (12.19) are a complete set of all constraints in the system, both primary and secondary. We will assume that they are independent of each other, meaning that removing any one of them would change the points (q,p) that satisfy all the constraints. The details of making the constraints independent were given in Chapter V when obtaining the formulas (5.12).

We have also shown that it is possible to get the correct Hamiltonian by using the primary constraints only, but it seems that it will give no significant advantage. Also, it could be much more complicated, since then we cannot easily use the velocities from (12.10) to eliminate the velocities from the pre-Hamiltonian K.

CHAPTER XIII

THE DIRAC'S BRACKETS

In this chapter, we will explain why Dirac's Brackets, designed as a replacement for the Poisson Brackets, are needed. We will also describe how Dirac's Brackets are constructed and list their basic properties.

1. The Need for Modifications of the Poisson Brackets

A reader who finished Chapter XII may conclude that we do not need anything else. We started from a Lagrangian, and described how, in principle, to obtain a Hamiltonian and the Poisson Brackets that reproduce the Euler-Lagrange equations of motion as their Hamilton's equations. So, in a sense, we have finished the process.

However, for a physicist, the work is not finished. A physicist would think about the next step, which is the quantization of the system. Quantization is a large subject and definitely out of the scope of this book. However, just to understand the basic motivation behind the introduction of the Dirac's Brackets, we need to provide some information here.

In the process of quantization, the physical variables q_j, p_j, $j = 1, ..., N$, are replaced by linear operators (which may be thought of as infinitely dimensional matrices) $\hat{q}_j, \hat{p}_j$ acting on infinitely dimensional vectors. The replacement is such that the basic Poisson Bracket, namely

$$\{q_j, p_j\} = 1 \quad \text{(no summation over } j\text{)}, \tag{13.1}$$

is replicated by the linear operators satisfying

$$\left[\hat{q}_j, \hat{p}_j\right] = i\hbar \hat{1} \quad \text{(no summation over } j\text{)}, \tag{13.2}$$

where the commutator $\left[\hat{q}_j, \hat{p}_j\right]$ is defined as:

$$[\hat{q}_j, \hat{p}_j] = \hat{q}_j \hat{p}_j - \hat{p}_j \hat{q}_j \text{ (no summation over } j\text{)}, \qquad (13.3)$$

i is the basic imaginary number $i = \sqrt{-1}$, $\hat{1}$ is the infinitely dimensional identity matrix, and $\hbar$ is a basic quantity of quantum physics, called the reduced Planck's constant. The constant $\hbar$ is also called the "crossed-h," or "h-bar," or the Dirac's constant.

The basic problem with the constraints is the following:

Assume that we have two constraints $q_1 = 0, p_1 = 0$. We then would expect that after the quantization we have $\hat{q}_1 = \hat{0}, \hat{p}_1 = \hat{0}$, where $\hat{0}$ is the zero operator (a matrix with all entries equal to zero), so that the results we obtain after the quantization would agree with the classical results. But then, we will also get:

$$[\hat{q}_1, \hat{p}_1] = \hat{q}_1 \hat{p}_1 - \hat{p}_1 \hat{q}_1 = \hat{0}\hat{0} - \hat{0}\hat{0} = \hat{0} \neq i\hbar \hat{1}.$$

So, the basic condition (13.2) of the quantization could not be satisfied.

The problem can be resolved in two different ways. One way would be to try to modify the rules by which the Poisson Brackets are quantized. However, this would mean changing the basic rules of quantization, which does not seem to be very feasible. The second way, proposed by Dirac, would be to replace the Poisson Brackets with different brackets.

These new dynamical brackets, typically denoted by $\{F, G\}_D$ and called "Dirac's Brackets," have the same basic properties as the Poisson Brackets. They give the same Hamilton's equations of motion as the Poisson Brackets but, differently from the Poisson Brackets, they also give:

$$\{q_1, p_1\}_D = 0, \qquad (13.4)$$

in the case we have the constraints $q_1 = 0, p_1 = 0$ in our system. So then, when quantizing, we have the requirement:

$$[\hat{q}_1, \hat{p}_1] = i\hbar\hat{0} = \hat{0}, \tag{13.5}$$

instead of (13.2), and this condition can be easily satisfied by the quantization $\hat{q}_1 = \hat{0}, \hat{p}_1 = \hat{0}$.

2. Separating the Variables into Two Classes

Let us say we have a system described by the variables (q_i, p_i), $i = 1,...,N$, and by the equations of motion (12.19) in Chapter XII.

The last line of the equations (12.19) are the constraints:

$$\varphi_s(q, p) = 0, \quad s = 1,...,S. \tag{13.6}$$

The constraints in (13.6) are all constraints in the system, both primary and secondary. As usual, we assume that they are independent, meaning that removing any one of them would change the points (q, p), which satisfy all the constraints in the system (13.6). The details of making the constraints independent were given in Chapter V when obtaining the formulas (5.12).

Recall that by a variable we understand any function $F = F(q, p)$. The functions used in the expressions defining the constraints (13.6) are also variables.

Now we divide all variables into two Classes.

1) Class I variables.

By definition, a variable $F = F(q, p)$ is of Class I (or first-class) if, for any constraint in (13.6), we have:

$$\{F, \varphi_s\} = 0, \tag{13.7}$$

on all the (q, p) satisfying all constraints in (13.6).

The brackets in (13.7) are the usual Poisson Brackets.

Notice that while φ_s and F are defined for all (q, p), the equation (13.7) is supposed to hold only for (q, p) that satisfy the constraints (13.6).

2) Class II variables.

A variable $F = F(q, p)$ is of Class II, (or second-class), if it is not of Class I.

More directly, $F = F(q, p)$ is of Class II, if there exists at least one point (q, p) that satisfies all the constraints (13.6), and there exists at least one constraint, say φ_s, such that we have:

$$\{F, \varphi_s\} \neq 0, \text{ for the constraint at the point.} \tag{13.8}$$

Again, notice that while φ_s and F are defined for all (q, p), the inequality (13.8) is supposed to hold only for some of the (q, p) that satisfy the constraints (13.6).

3. Constructing a Minimal Complete Set of Class II Constraints

Now, let us present the procedure to make the decision on which constraints will be used for defining the Dirac's Brackets. Dirac's Brackets use only the constraints which are "really" of the Class II.

Constraints φ_s in (13.6), when looked at as functions of all (q, p), are variables. Therefore, as variables, the constraints can also be classified as belonging to either Class I or Class II.

However, dividing sets of constraints into Class I and Class II subsets is somewhat arbitrary. For example, say we have a set of constraints:

$$\begin{aligned} \chi_1 &= q_1 = 0, \\ \chi_2 &= p_1 = 0, \\ \chi_3 &= q_1 + q_2 = 0. \end{aligned} \tag{13.9}$$

For the constraints in (13.9) we have $\{\chi_1, \chi_2\} = 1$, and $\{\chi_3, \chi_2\} = 1$. So, all three constraints in

(13.9) are of Class II.

Now, let us look at a different set of constraints:

$$\chi_1 = q_1 = 0,$$
$$\chi_2 = p_1 = 0, \quad (13.10)$$
$$\chi_3 = q_2 = 0.$$

Here we have $\{\chi_1, \chi_3\} = 0$, and $\{\chi_2, \chi_3\} = 0$. So, the constraint χ_3 is of Class I.

However, it is easy to see that sets of constraints (13.9) and (13.10) create identical restrictions on possible (q, p). In this aspect, the two sets of constraints are completely equivalent, and either one can be used. Still, the individual constraints in these sets are very different in terms of belonging to Class I or Class II.

What we are going to do now is to show how to modify a given set of constrains to replace it with a set of constraints that gives the same restrictions on (q, p), but has as many constraints moved from Class II to Class I as possible.

Let us start with a set (13.6) of constraints. Then, in this set let us denote the constraints that belong to Class I, by:

$$\varphi_a = \varphi_a(q, p), \quad a = 1, \ldots, A. \quad (13.11)$$

The other constraints in (13.6) are of Class II. We will denote them by:

$$\chi_b = \chi_b(q, p), \quad b = 1, \ldots, B. \quad (13.12)$$

Now consider a different set of functions, namely:

$$\eta_c = \eta_c(q,p) = \sum_{b=1}^{B} u_{cb}(q,p) \cdot \chi_b(q,p), \quad c = 1,\ldots,B. \tag{13.13}$$

Where $u_{cb}(q,p)$ are such functions of all (q,p) (not just the (q,p) satisfying the constraints), that at any given (q,p), the matrix:

$$X_{cb} = u_{cb}(q,p), \quad b,c = 1,\ldots,B, \tag{13.14}$$

is invertible. Invertibility of X_{cb} assures that the functions $\eta_c(q,p)$ are all equal to zero at exactly the same points (q,p) at which all $\chi_b(q,p)$ are equal to zero. So, the set of constraints:

$$\eta_c = \eta_c(q,p) = 0, \quad c = 1,\ldots,B, \tag{13.15}$$

can replace the original constraints (13.12), in the sense that it produces the same restrictions on (q,p).

Moreover, the results of Chapter VI tell us that any possible set of functions that could be used to describe the same set of (q,p) as the original constraints (13.12) can be expressed as (13.13). In other words, the formulas (13.13) cover all possible sets of constraints that are equivalent to the set of constraints (13.12), in the sense that they produce the same restrictions on (q,p).

Notice that changing the constraints from (13.12) to (13.13) is not changing which variables belong to Class I, since if F is of Class I, then we have:

$$\begin{aligned}\{F,\eta_c\} &= \left\{F, \sum_{b=1}^{B} u_{cb}(q,p) \cdot \chi_b(q,p)\right\} = \sum_{b=1}^{B}\{F, u_{cb}(q,p) \cdot \chi_b(q,p)\} = \\ &= \sum_{b=1}^{B}\left[u_{cb}(q,p) \cdot \{F, \chi_b(q,p)\} + \{F, u_{cb}(q,p)\} \cdot \chi_b(q,p)\right] = \\ &= \sum_{b=1}^{B}\left[u_{cb}(q,p) \cdot 0 + \{F, u_{cb}(q,p)\} \cdot 0\right] = 0,\end{aligned} \tag{13.16}$$

for all (q,p) satisfying all constraints (13.11), (13.12), and (13.13).

Notice that no matter what functions $u_{cb}(q,p)$ we choose in (13.13), any Class I variable must have zero Poisson Brackets with all constraints written in the form of (13.13). This is the key to identify constraints expressed in the form (13.13) that are of Class I.

Consider a single constraint written as:

$$\eta = \eta(q,p) = \sum_{d=1}^{B} w_d(q,p) \cdot \chi_d(q,p). \tag{13.17}$$

Assume for a moment that the constraint (13.17) belongs to Class I. Then it must give Poison Bracket equal to zero with any single constraint that belongs to (13.13). Let us write a general single constraint from the set (13.13) as:

$$\gamma = \gamma(q,p) = \sum_{b=1}^{B} u_b(q,p) \cdot \chi_b(q,p). \tag{13.18}$$

When we look at formulas (13.17) and (13.18), they look essentially identical. Yet, we think about them differently. We want to find just one η expressed by (13.17) that will give zero Poisson Brackets with all possible constraints γ in (13.18).

Let us now calculate the Poisson Bracket:

$$\{\gamma,\eta\} = \left\{ \sum_{b=1}^{B} u_b(q,p) \cdot \chi_b(q,p), \sum_{d=1}^{B} w_d(q,p) \cdot \chi_d(q,p) \right\} =$$

(In the continuation below, we drop (q,p) to make the expressions shorter.)

$$= \sum_{b=1}^{B}\sum_{d=1}^{B} \{u_b \cdot \chi_b, w_d \cdot \chi_d\} = \sum_{b=1}^{B}\sum_{d=1}^{B} \left[u_b\{\chi_b, w_d \cdot \chi_d\} + \chi_b\{u_b, w_d \cdot \chi_d\} \right] =$$

(continued on the next page)

$$= \sum_{b=1}^{B}\sum_{d=1}^{B}\left[u_b\{\chi_b, w_d \cdot \chi_d\} + 0\cdot\{u_b, w_d\cdot\chi_d\}\right] = \sum_{b=1}^{B}\sum_{d=1}^{B}\left[u_b\{\chi_b, w_d\cdot\chi_d\}\right] =$$

$$= \sum_{b=1}^{B}\sum_{d=1}^{B}\left[u_b\{\chi_b,\chi_d\}w_d + u_b\{\chi_b,w_d\}\chi_d\right] = \sum_{b=1}^{B}\sum_{d=1}^{B}\left[u_b\{\chi_b,\chi_d\}w_d + u_b\{\chi_b,w_d\}\cdot 0\right] =$$

$$= \sum_{b=1}^{B}\sum_{d=1}^{B}\left[u_b\{\chi_b,\chi_d\}w_d\right]$$

So, we have:

$$\{\gamma,\eta\} = \sum_{b=1}^{B}\sum_{d=1}^{B}\left[u_b\{\chi_b,\chi_d\}w_d\right]. \tag{13.19}$$

The formula (13.19) can be written in a matrix form as:

$$\{\gamma,\eta\} = (u_1, u_2, ..., u_B)\begin{pmatrix} \{\chi_1,\chi_1\} & \{\chi_1,\chi_2\} & \cdots & \{\chi_1,\chi_B\} \\ \{\chi_2,\chi_1\} & \vdots & \vdots & \{\chi_2,\chi_B\} \\ \vdots & \vdots & \vdots & \vdots \\ \{\chi_B,\chi_1\} & \{\chi_B,\chi_2\} & \cdots & \{\chi_B,\chi_B\} \end{pmatrix}\begin{pmatrix} w_1 \\ w_2 \\ \vdots \\ w_B \end{pmatrix}. \tag{13.20}$$

Recall that we use formula (13.20) to see in which class, Class I or Class II, the constraint η is. Since each class is defined using only the points (q, p) that satisfy all constraints, (13.11), (13.12), and (13.13), we do the check below for such points (q, p) only.

For each such point we have two possibilities about the matrix in the middle of the formula (13.20):

1) The matrix is invertible.

In this case, from basic linear algebra, we know that if the vector $\begin{pmatrix} w_1 \\ w_2 \\ \vdots \\ w_B \end{pmatrix}$ is non-zero (not all entries are zero), then also the vector $\begin{pmatrix} \{\chi_1, \chi_1\} & \{\chi_1, \chi_2\} & \cdots & \{\chi_1, \chi_B\} \\ \{\chi_2, \chi_1\} & \vdots & \vdots & \{\chi_2, \chi_B\} \\ \vdots & \vdots & \vdots & \vdots \\ \{\chi_B, \chi_1\} & \{\chi_B, \chi_2\} & \cdots & \{\chi_B, \chi_B\} \end{pmatrix} \begin{pmatrix} w_1 \\ w_2 \\ \vdots \\ w_B \end{pmatrix}$ must be non-zero. Then we can easily find such vector $(u_1, u_2, \ldots, u_B)$ that will give a non-zero result in (13.20).

This means that no matter what constraint η we choose, we will be able to find such γ that $\{\gamma, \eta\} \neq 0$. But this means that the set (13.12) is made of constraints of Class II. No constraint can be created from constraints in (13.12) that would be of Class I.

So, when the matrix in (13.20) is invertible, we have finished the process. We have the set of constraints (13.12) as the set of constraints which are all of Class II, and this set cannot be made smaller.

2. The second possibility is that the matrix in (13.20) is not invertible. Then, again from the basic linear algebra, we know that there must be such a non-zero vector $\begin{pmatrix} w_1 \\ w_2 \\ \vdots \\ w_B \end{pmatrix}$ that

$$\begin{pmatrix} \{\chi_1, \chi_1\} & \{\chi_1, \chi_2\} & \cdots & \{\chi_1, \chi_B\} \\ \{\chi_2, \chi_1\} & \vdots & \vdots & \{\chi_2, \chi_B\} \\ \vdots & \vdots & \vdots & \vdots \\ \{\chi_B, \chi_1\} & \{\chi_B, \chi_2\} & \cdots & \{\chi_B, \chi_B\} \end{pmatrix} \begin{pmatrix} w_1 \\ w_2 \\ \vdots \\ w_B \end{pmatrix} = \begin{pmatrix} 0 \\ 0 \\ \vdots \\ 0 \end{pmatrix}.$$

Then, placing the vector $\begin{pmatrix} w_1 \\ w_2 \\ \vdots \\ w_B \end{pmatrix}$ in (13.20) will give us $\{\gamma, \eta\} = 0$ no matter what γ we choose.

This means that the constraint η is of Class I.

So, we include the η into Class I constraints described in (13.11). Usually, at this point, we need to modify our Class II constraints in (13.12), since the constraint η is not one of the constraints in

(13.11). This modification of the constraints in (13.12) can be done by defining new constraints using (13.13) in such a way that $\eta = \eta_B$. Then, we just drop the η_B from the set of constraints.

The process of identifying and removing constraints can now be repeated until we get an invertible matrix. Then, we have no more constraints of Class I, and we are done. The set of Class II constraints we obtain is called a minimal complete set of Class II constraints.

At the end of it, we have the following situation:

We have a set of constraints of Class I. We will denote them as:

$$\varphi_a = \varphi_a(q,p), \quad a = 1,\ldots,A. \tag{13.20}$$

Notice that, in general, the number A in (13.20) is larger than A in (13.11). However, we decided to use the same symbol for it.

We also have a minimal complete set of Class II constraints, denoted by

$$\chi_b = \chi_b(q,p), \quad b = 1,\ldots,B, \tag{13.21}$$

which has the property that the constraint matrix $C^{-1} = C^{-1}(q,p)$, defined as:

$$C^{-1} = \begin{pmatrix} \{\chi_1,\chi_1\} & \{\chi_1,\chi_2\} & \cdots & \{\chi_1,\chi_B\} \\ \{\chi_2,\chi_1\} & \vdots & \vdots & \{\chi_2,\chi_B\} \\ \vdots & \vdots & \vdots & \vdots \\ \{\chi_B,\chi_1\} & \{\chi_B,\chi_2\} & \cdots & \{\chi_B,\chi_B\} \end{pmatrix}, \tag{13.22}$$

is invertible. The inverse of matrix C^{-1} will be denoted by C.

It may look a little strange that we use the symbol C^{-1} for the newly defined matrix, and then C for its inverse, but we decided to use the notation consistent with what Dirac used in his original work.

As mentioned before, the matrix C^{-1} is defined for (q,p) that satisfy all constraints of the system, both Class I and Class II. Notice that invertibility of C^{-1} may change from one point (q,p) to the next, so our considerations are local.

Notice that because of the properties of the Poisson Brackets, the matrix C^{-1} is antisymmetric. Since it is invertible, it must be of even dimension, since the basic linear algebra tells us that antisymmetric invertible matrices must be of even dimensions. This tells us that B in (13.21) is even, so the minimal complete set of Class II constraints can be written as:

$$\chi_b = \chi_b(q,p), \quad b=1,\ldots,2K. \tag{13.23}$$

4. Definition of the Dirac's Brackets

Having a system with the Poisson Brackets $\{F,G\}$, a minimal complete system of Class II constraints (13.22) with the constraints matrix C^{-1}, we define the Dirac's Brackets $\{F,G\}_D$ as:

$$\{F,G\}_D = \{F,G\} - \sum_{b=1}^{B}\sum_{c=1}^{B}\{F,\chi_b\}\cdot C_{bc} \cdot \{\chi_c,G\}, \tag{13.24}$$

where the matrix $C_{bc} = C_{bc}(q,p)$ is the inverse of the matrix C^{-1} from (13.22).

Let us notice that for given functions F and G, the Dirac Bracket $\{F,G\}_D$ is also a function of (q,p). The functions F and G, as well as χ_b's in (13.24) are defined for all (q,p). Since C^{-1} is invertible for a point (q,p), and because of continuity reasons (the non-zero determinant is a continuous function of the entries), it is also invertible in an open neighborhood of the point. So, the Dirac Brackets are defined in open neighborhoods of each (q,p) that satisfies all the constraints.

Therefore $\dfrac{\partial \{F,G\}_D}{\partial q_i}$ and $\dfrac{\partial \{F,G\}_D}{\partial p_i}$, $i=1,...,N$, are well defined in the sense that the constraints do not interfere with infinitesimal change in one variable while keeping other variables constant, as required for partial derivatives. This is because Dirac's Brackets are defined not only for the points that satisfy the constraints but also the points in their open neighborhoods.

Since the above derivatives are well defined, then also expressions containing multiple Dirac Brackets, for example $\{\{F,G\}_D, E\}_D$, where E, F, and G are some functions of $(q_1, q_2, ..., q_N, p_1, p_2, ..., p_N)$, are well-defined as well.

To keep things simple, and not repeat the sentence about open neighborhoods again and again, for the remainder of this text, we will assume that the Dirac Brackets are defined for all (q,p). However, this fact should be looked at on a case-by-case basis for each specific situation.

5. Basic Properties of the Dirac's Brackets

Below is the list of basic properties of the Dirac's Brackets. They are identical to the properties of the Poisson Brackets proved in Chapter VII. We will list them here for completeness. We will not prove them though, since their proof becomes obvious to everyone who will read Chapter XIV of this book.

Basic properties of the Dirac's Brackets are:

If E, F, and G are some functions of $(q_1, q_2, ..., q_N, p_1, p_2, ..., p_N)$, and a and b are constants, then we have:

a) $\{F,G\}_D = -\{G,F\}_D$ \hfill (13.25)

b) $\{aF,G\}_D = a\{F,G\}_D$ \hfill (13.26)

c) $\{F,aG\}_D = a\{F,G\}_D$ \hfill (13.27)

d) $\{F+G, E\}_D = \{F, E\}_D + \{G, E\}_D$ \hfill (13.28)

e) $\{F, G+E\}_D = \{F, G\}_D + \{F, E\}_D$ \hfill (13.29)

f) $\{aF + bG, E\}_D = a\{F, E\}_D + b\{G, E\}_D$ \hfill (13.30)

h) $\{F, aG + bE\}_D = a\{F, G\}_D + b\{F, E\}_D$ \hfill (13.31)

i) $\{FG, E\}_D = F\{G, E\}_D + G\{F, E\}_D$ \hfill (13.32)

j) $\{F, GE\}_D = G\{F, E\}_D + E\{F, G\}_D$ \hfill (13.33)

h) Jacobi Identity (notice the cyclical nature of the rotation of functions)

$$\{\{F, G\}_D, E\}_D + \{\{G, E\}_D, F\}_D + \{\{E, F\}_D, G\}_D = 0 \quad (13.34)$$

Instead of showing here that the properties (13.25) to (13.34) are true, we will wait until Chapter XIV, where we will present the interpretation of the Dirac's Brackets, which will make the properties obvious. Right now, we are just going to use them until the end of this chapter.

6. Special Properties of the Dirac's Brackets

Now we go to more advanced properties of Dirac's Brackets:

A) If we denote all final constraints of Class I by φ_a and all final constraints of Class II by χ_b, then we have:

$$\{\varphi_a, \varphi_d\}_D = \{\varphi_a, \varphi_d\} - \sum_{b=1}^{B}\sum_{c=1}^{B} \{\varphi_a, \chi_b\} \cdot C_{bc} \cdot \{\chi_c, \varphi_d\} =$$
$$= 0 - \sum_{b=1}^{B}\sum_{c=1}^{B} 0 \cdot C_{bc} \cdot 0 = 0. \quad (13.35)$$

We also have:

$$\{\varphi_a, \chi_d\}_D = \{\varphi_a, \chi_d\} - \sum_{b=1}^{B}\sum_{c=1}^{B}\{\varphi_a, \chi_b\} \cdot C_{bc} \cdot \{\chi_c, \chi_d\} =$$
$$= 0 - \sum_{b=1}^{B}\sum_{c=1}^{B} 0 \cdot C_{bc} \cdot \{\chi_c, \chi_d\} = 0.$$

(13.36)

Finally, we have:

$$\{\chi_a, \chi_d\}_D = \{\chi_a, \chi_d\} - \sum_{b=1}^{B}\sum_{c=1}^{B}\{\chi_a, \chi_b\} \cdot C_{bc} \cdot \{\chi_c, \chi_d\} =$$
$$= \{\chi_a, \chi_d\} - \sum_{b=1}^{B}\sum_{c=1}^{B}\{\chi_a, \chi_b\} \cdot C_{bc} \cdot \left(C^{-1}\right)_{cd} =$$
$$= \{\chi_a, \chi_d\} - \sum_{b=1}^{B}\{\chi_a, \chi_b\} \cdot \delta_{bd} = \{\chi_a, \chi_d\} - \{\chi_a, \chi_d\} = 0.$$

(13.37)

This means that all constraints have zero Dirac's Brackets with all other constraints.

B) The Dirac's Bracket of the Hamiltonian with any constraint of the system is equal to zero.

Again, denote all constraints of Class I by φ_a and all final constraints of Class II by χ_b. Denote the Hamiltonian by H. Then, using the Hamilton's equations (12.19), we have:

$$\dot{\varphi}_a = \{\varphi_a, H\} = 0,$$
$$\dot{\chi}_b = \{\chi_b, H\} = 0.$$

(13.38)

The above is true because the constraints are supposed to hold in time, so their time derivatives are equal to zero. Then:

$$\{\varphi_a, H\}_D = \{\varphi_a, H\} - \sum_{b=1}^{B}\sum_{c=1}^{B}\{\varphi_a, \chi_b\} \cdot C_{bc} \cdot \{\chi_c, H\} =$$
$$= 0 - \sum_{b=1}^{B}\sum_{c=1}^{B}\{\varphi_a, \chi_b\} \cdot C_{bc} \cdot 0 = 0 - 0 = 0,$$

(13.39)

$$\{\chi_d, H\}_D = \{\chi_d, H\} - \sum_{b=1}^{B}\sum_{c=1}^{B} \{\chi_d, \chi_b\} \cdot C_{bc} \cdot \{\chi_c, H\} =$$
$$= 0 - \sum_{b=1}^{B}\sum_{c=1}^{B} \{\chi_d, \chi_b\} \cdot C_{bc} \cdot 0 = 0 - 0 = 0.$$
(13.40)

C) The Dirac's Bracket of any variable F with any Class II constraint that was used to define the Dirac's Brackets is equal to zero.

We have:

$$\{\chi_d, F\}_D = \{\chi_d, F\} - \sum_{b=1}^{B}\sum_{c=1}^{B} \{\chi_d, \chi_b\} \cdot C_{bc} \cdot \{\chi_c, F\} =$$
$$= \{\chi_d, F\} - \sum_{b=1}^{B}\sum_{c=1}^{B} (C^{-1})_{db} \cdot C_{bc} \cdot \{\chi_c, F\} =$$
$$= \{\chi_d, F\} - \sum_{c=1}^{B} \delta_{dc} \{\chi_c, F\} = \{\chi_d, F\} - \{\chi_d, F\} = 0.$$
(13.41)

D) The Dirac's Bracket of any variable F with the Hamiltonian H is equal to the Poisson Bracket.

We have:

$$\{F, H\}_D = \{F, H\} - \sum_{b=1}^{B}\sum_{c=1}^{B} \{F, \chi_b\} \cdot C_{bc} \cdot \{\chi_c, H\} =$$
$$= \{F, H\} - \sum_{b=1}^{B}\sum_{c=1}^{B} \{F, \chi_b\} \cdot C_{bc} \cdot 0 = \{F, H\}.$$
(13.42)

E) The Dirac's Bracket can replace the Poisson Bracket in the Hamilton's equations of motion.

We have:

$$\dot{F} = \{F, H\} = \{F, H\}_D.$$

Therefore:

$$\dot{F} = \{F, H\}_D. \tag{13.43}$$

F) The Class II constraints that were used to define the Dirac's Brackets can be used to do substitutions in the Hamiltonian, without changing the Hamilton's equations, provided we use the Dirac's Brackets in the Hamilton's equations, not the usual Poisson Brackets or partial derivatives.

From Chapter VI, we know that general substitutions in the Hamiltonian using the Class II constraints $\chi_b = \chi_b(q, p)$, $b = 1, ..., B$, can be given as:

$$H_{SUB} = H + \sum_{b=1}^{B} u_b \cdot \chi_b. \tag{13.44}$$

Then we have:

$$\begin{aligned}
\{F, H_{SUB}\}_D &= \left\{F, H + \sum_{b=1}^{B} u_b \cdot \chi_b\right\}_D = \{F, H\}_D + \left\{F, \sum_{b=1}^{B} u_b \cdot \chi_b\right\}_D = \\
&= \{F, H\}_D + \sum_{b=1}^{B} \{F, u_b \cdot \chi_b\}_D = \\
&= \{F, H\}_D + \sum_{b=1}^{B} u_b \cdot \{F, \chi_b\}_D + \sum_{b=1}^{B} \chi_b \cdot \{F, u_b\}_D = \\
&= \{F, H\}_D + \sum_{b=1}^{B} u_b \cdot 0 + \sum_{b=1}^{B} 0 \cdot \{F, u_b\}_D = \{F, H\}_D + 0 + 0 = \{F, H\}_D.
\end{aligned} \tag{13.45}$$

Then we have:

$$\dot{F} = \{F, H\}_D = \{F, H_{SUB}\}_D, \tag{13.46}$$

and finally:

$$\dot{F} = \{F, H_{SUB}\}_D. \tag{13.47}$$

In practice, we usually use the symbol H for H_{SUB}.

It may be useful to know that we can use the Class II constraints that were used to define the Dirac's Brackets to simplify the Hamiltonian. At the same time, it is important to remember that, in general, we are not allowed to use the Class I constraints to modify the Hamiltonian.

G) All constraints of the system can be used to do substitutions in any variable that is not going to be used as a Hamiltonian, without changing the Hamilton's equations for that variable.

From Chapter VI, we know that general substitutions in a variable F, using Class I constraints $\varphi_a = \varphi_a(q,p)$, $a=1,...,A$, and Class II constraints $\chi_b = \chi_b(q,p)$, $b=1,...,B$, can, in general, be given as:

$$F_{SUB} = F + \sum_{a=1}^{A} u_a \cdot \varphi_a + \sum_{b=1}^{B} w_b \cdot \chi_b. \tag{13.48}$$

Then we have:

$$\{F_{SUB}, H\}_D = \left\{F + \sum_{a=1}^{A} u_a \cdot \varphi_a + \sum_{b=1}^{B} w_b \cdot \chi_b, H\right\}_D =$$

$$= \{F, H\}_D + \left\{\sum_{a=1}^{A} u_a \cdot \varphi_a, H\right\}_D + \left\{\sum_{b=1}^{B} w_b \cdot \chi_b, H\right\}_D =$$

$$= \{F, H\}_D + \sum_{a=1}^{A} \{u_a \cdot \varphi_a, H\}_D + \sum_{b=1}^{B} \{w_b \cdot \chi_b, H\}_D =$$

$$= \{F, H\}_D + \sum_{a=1}^{A} u_a \cdot \{\varphi_a, H\}_D + \sum_{a=1}^{A} \varphi_a \cdot \{u_a, H\}_D +$$

$$+ \sum_{b=1}^{B} w_b \cdot \{\chi_b, H\}_D + \sum_{b=1}^{B} \chi_b \cdot \{w_b, H\}_D =$$

$$= \{F, H\}_D + \sum_{a=1}^{A} u_a \cdot 0 + \sum_{a=1}^{A} 0 \cdot \{u_a, H\}_D +$$

$$+ \sum_{b=1}^{B} w_b \cdot 0 + \sum_{b=1}^{B} 0 \cdot \{w_b, H\}_D = \{F, H\}_D. \tag{13.49}$$

So, we have:

$$\{F_{SUB}, H\}_D = \{F, H\}_D. \tag{13.50}$$

Then, we also have:

$$\dot{F}_{SUB} = \{F_{SUB}, H\}_D = \{F, H\}_D = \dot{F}. \tag{13.51}$$

H) General formula for Dirac's Brackets

Using the basic formula (13.24), namely $\{F, G\}_D = \{F, G\} - \sum_{b=1}^{B}\sum_{c=1}^{B}\{F, \chi_b\} \cdot C_{bc} \cdot \{\chi_c, G\}$, may not be very convenient when studying a specific application. It may be worth modifying it.

Let us change the formula (13.24) to the matrix form. We split the Dirac formula (13.24) into smaller parts, introducing symbols for parts of it. We start by introducing new matrix symbols:

$$\left[\frac{\partial F}{\partial q}, \frac{\partial F}{\partial p}\right] = \left(\frac{\partial F}{\partial q_1}, ..., \frac{\partial F}{\partial q_N}, \frac{\partial F}{\partial p_1}, ..., \frac{\partial F}{\partial p_N}\right), \tag{13.52}$$

$$[J] = \left(\begin{array}{cccc|cccc} 0 & \cdots & \cdots & 0 & 1 & 0 & \cdots & 0 \\ \vdots & \ddots & & \vdots & 0 & \ddots & & \vdots \\ \vdots & & \ddots & \vdots & \vdots & & \ddots & 0 \\ 0 & \cdots & \cdots & 0 & 0 & \cdots & \cdots & 1 \\ \hline -1 & 0 & \cdots & 0 & 0 & \cdots & \cdots & 0 \\ 0 & \ddots & & \vdots & \vdots & \ddots & & \vdots \\ \vdots & & \ddots & 0 & \vdots & & \ddots & \vdots \\ 0 & \cdots & \cdots & -1 & 0 & \cdots & \cdots & 0 \end{array}\right), \tag{13.53}$$

$$\left[\frac{\partial G}{\partial q} \atop \frac{\partial G}{\partial p}\right] = \begin{pmatrix} \frac{\partial G}{\partial q_1} \\ \vdots \\ \frac{\partial G}{\partial q_N} \\ \frac{\partial G}{\partial p_1} \\ \vdots \\ \frac{\partial G}{\partial p_N} \end{pmatrix} \qquad (13.54)$$

$$\left[\frac{\partial \chi}{\partial q} \atop \frac{\partial \chi}{\partial p}\right] = \begin{pmatrix} \frac{\partial \chi_1}{\partial q_1} & \frac{\partial \chi_2}{\partial q_1} & \cdots & \frac{\partial \chi_B}{\partial q_1} \\ \vdots & \vdots & & \vdots \\ \frac{\partial \chi_1}{\partial q_N} & \frac{\partial \chi_2}{\partial q_N} & \cdots & \frac{\partial \chi_B}{\partial q_N} \\ \frac{\partial \chi_1}{\partial p_1} & \frac{\partial \chi_2}{\partial p_1} & \cdots & \frac{\partial \chi_B}{\partial p_1} \\ \vdots & \vdots & & \vdots \\ \frac{\partial \chi_1}{\partial p_N} & \frac{\partial \chi_2}{\partial p_N} & \vdots & \frac{\partial \chi_B}{\partial p_N} \end{pmatrix}, \qquad (13.55)$$

$$\left[\frac{\partial \chi}{\partial q}, \frac{\partial \chi}{\partial p}\right] = \begin{pmatrix} \frac{\partial \chi_1}{\partial q_1} & \cdots & \frac{\partial \chi_1}{\partial q_N} & \frac{\partial \chi_1}{\partial p_1} & \cdots & \frac{\partial \chi_1}{\partial p_N} \\ \frac{\partial \chi_2}{\partial q_1} & \cdots & \frac{\partial \chi_2}{\partial q_N} & \frac{\partial \chi_2}{\partial p_1} & \cdots & \frac{\partial \chi_2}{\partial p_N} \\ \vdots & & \vdots & \vdots & & \vdots \\ \frac{\partial \chi_B}{\partial q_1} & \cdots & \frac{\partial \chi_B}{\partial q_N} & \frac{\partial \chi_B}{\partial p_1} & \cdots & \frac{\partial \chi_N}{\partial p_N} \end{pmatrix}, \qquad (13.56)$$

$$[C] = C_{bc}. \qquad (13.57)$$

Using the matrices defined above, we have:

$$\{F,G\} = \left[\frac{\partial F}{\partial q}, \frac{\partial F}{\partial p}\right][J]\begin{bmatrix}\frac{\partial G}{\partial q} \\ \frac{\partial G}{\partial p}\end{bmatrix}, \tag{13.58}$$

$$\sum_{b=1}^{B}\sum_{c=1}^{B}\{F,\chi_b\}\cdot C_{bc}\cdot\{\chi_c,G\} =$$

$$= \left[\frac{\partial F}{\partial q}, \frac{\partial F}{\partial p}\right][J]\begin{bmatrix}\frac{\partial \chi}{\partial q} \\ \frac{\partial \chi}{\partial p}\end{bmatrix}[C]\left[\frac{\partial \chi}{\partial q}, \frac{\partial \chi}{\partial p}\right][J]\begin{bmatrix}\frac{\partial G}{\partial q} \\ \frac{\partial G}{\partial p}\end{bmatrix}. \tag{13.59}$$

Therefore, the Dirac's Brackets are expressed as:

$$\{F,G\}_D = \{F,G\} - \sum_{b=1}^{B}\sum_{c=1}^{B}\{F,\chi_b\}\cdot C_{bc}\cdot\{\chi_c,G\} =$$

$$= \left[\frac{\partial F}{\partial q}, \frac{\partial F}{\partial p}\right][J]\begin{bmatrix}\frac{\partial G}{\partial q} \\ \frac{\partial G}{\partial p}\end{bmatrix} - \left[\frac{\partial F}{\partial q}, \frac{\partial F}{\partial p}\right][J]\begin{bmatrix}\frac{\partial \chi}{\partial q} \\ \frac{\partial \chi}{\partial p}\end{bmatrix}[C]\left[\frac{\partial \chi}{\partial q}, \frac{\partial \chi}{\partial p}\right][J]\begin{bmatrix}\frac{\partial G}{\partial q} \\ \frac{\partial G}{\partial p}\end{bmatrix} = \tag{13.60}$$

$$= \left[\frac{\partial F}{\partial q}, \frac{\partial F}{\partial p}\right]\left[[J] - [J]\begin{bmatrix}\frac{\partial \chi}{\partial q} \\ \frac{\partial \chi}{\partial p}\end{bmatrix}[C]\left[\frac{\partial \chi}{\partial q}, \frac{\partial \chi}{\partial p}\right][J]\right]\begin{bmatrix}\frac{\partial G}{\partial q} \\ \frac{\partial G}{\partial p}\end{bmatrix}.$$

Introducing matrix $[D]$ as:

$$[D] = [J] - [J]\begin{bmatrix}\frac{\partial \chi}{\partial q} \\ \frac{\partial \chi}{\partial p}\end{bmatrix}[C]\left[\frac{\partial \chi}{\partial q}, \frac{\partial \chi}{\partial p}\right][J], \tag{13.61}$$

we get:

$$\{F,G\}_D = \left[\frac{\partial F}{\partial q}, \frac{\partial F}{\partial p}\right][D]\begin{bmatrix}\frac{\partial G}{\partial q} \\ \frac{\partial G}{\partial p}\end{bmatrix}. \tag{13.62}$$

Using (13.52) and (13.52), the Dirac's Brackets can be written as:

$$\{F,G\}_D = \left(\frac{\partial F}{\partial q_1}, ..., \frac{\partial F}{\partial q_N}, \frac{\partial F}{\partial p_1}, ..., \frac{\partial F}{\partial p_N}\right)[D]\begin{pmatrix}\frac{\partial G}{\partial q_1} \\ \vdots \\ \frac{\partial G}{\partial q_N} \\ \frac{\partial G}{\partial p_1} \\ \vdots \\ \frac{\partial G}{\partial p_N}\end{pmatrix}. \tag{13.63}$$

Using separate variables q_i and p_j in different combinations for F and G in (13.63), we will get a single 1 and multiple 0's in the vectors on both sides of the matrix $[D]$. Then multiplication will result in obtaining just one entry of matrix $[D]$. Comparing both sides will then give us:

$$[D] = \begin{pmatrix} \{q_1,q_1\}_D & \cdots & \{q_1,q_N\}_D & \{q_1,p_1\}_D & \cdots & \{q_1,p_N\}_D \\ \vdots & \ddots & \vdots & \vdots & \vdots & \vdots \\ \{q_N,q_1\}_D & \cdots & \{q_N,q_N\}_D & \{q_N,p_1\}_D & \cdots & \{q_N,p_N\}_D \\ \{p_1,q_1\}_D & \cdots & \{p_1,q_N\}_D & \{p_1,p_1\}_D & \cdots & \{p_1,p_N\}_D \\ \vdots & \vdots & \vdots & \vdots & \ddots & \vdots \\ \{p_N,q_1\}_D & \cdots & \{p_N,q_N\}_D & \{p_N,p_1\}_D & \cdots & \{p_N,p_N\}_D \end{pmatrix}. \tag{13.64}$$

So, in practice, if we need to calculate the matrix $[D]$, we can use either the matrix multiplication (13.61) or calculate each entry of (13.64) using the Dirac formula (13.24) directly. In both cases, the calculation is usually very tedious.

Using (13.64) and (13.63), we can also write the Dirac Brackets as:

$$\{F,G\}_D = \sum_{i=1}^{N}\sum_{j=1}^{N} \frac{\partial F}{\partial q_i} \cdot \{q_i, q_j\}_D \cdot \frac{\partial G}{\partial q_j} + \sum_{i=1}^{N}\sum_{j=1}^{N} \frac{\partial F}{\partial q_i} \cdot \{q_i, p_j\}_D \cdot \frac{\partial G}{\partial p_j} +$$
$$+ \sum_{i=1}^{N}\sum_{j=1}^{N} \frac{\partial F}{\partial p_i} \cdot \{p_i, q_j\}_D \cdot \frac{\partial G}{\partial q_j} + \sum_{i=1}^{N}\sum_{j=1}^{N} \frac{\partial F}{\partial p_i} \cdot \{p_i, p_j\}_D \cdot \frac{\partial G}{\partial p_j}.$$

(13.65)

7. The Dirac's Brackets Expressed in Different Variables

Starting with formula (13.63) and introducing the change of variables identical to (7.4), we obtain:

$$x_1 = x_1(q, p),$$
$$x_2 = x_2(q, p),$$
$$\vdots$$
$$x_{2N} = x_{2N}(q, p).$$

(13.66)

Repeating the steps between formulas (7.6) and (7.14), we get:

$$\{F,G\}_D = \left(\frac{\partial F}{\partial x_1}, \frac{\partial F}{\partial x_2}, \dots, \frac{\partial F}{\partial x_{2N}}\right) \begin{bmatrix} \frac{\partial x}{\partial q} & \frac{\partial x}{\partial p} \end{bmatrix} [D] \begin{bmatrix} \frac{\partial x}{\partial q} \\ \frac{\partial x}{\partial p} \end{bmatrix} \begin{pmatrix} \frac{\partial G}{\partial x_1} \\ \frac{\partial G}{\partial x_2} \\ \vdots \\ \frac{\partial G}{\partial x_{2N}} \end{pmatrix}.$$

(13.67)

Then, as before, using variables x_i and x_j, $i, j = 1, \dots, 2N$, for F and G in (13.67), we will get a single 1, and multiple $0's$ as the entries in the vectors on both sides of the matrix in the middle of (13.67). Then the matrix multiplication on the right side will result in obtaining just one entry of the matrix in the middle. Comparing both sides of (13.67) will then give:

$$\{F,G\}_D = \sum_{i=1}^{2N}\sum_{j=1}^{2N} \frac{\partial F}{\partial x_i} \cdot \{x_i, x_j\}_D \cdot \frac{\partial G}{\partial x_j}.$$

(13.68)

8. Obtaining the Newtonian Equations Using the Dirac's Brackets

Using the Hamilton's equations (13.43), we get the time derivatives of positions as:

$$\dot{q}_i = \{q_i, H\}_D. \tag{13.69}$$

The definitions of velocities are given by formulas (5.2) from Chapter V, namely:

$$v_i = \dot{q}_i. \tag{13.70}$$

So, we get:

$$v_i = \{q_i, H\}_D. \tag{13.71}$$

Then, we have:

$$\dot{v}_i = \{v_i, H\}_D. \tag{13.72}$$

So, using (13.71) in (13.72), we get:

$$\dot{v}_i = \{\{q_i, H\}_D, H\}_D. \tag{13.73}$$

The formulas (13.73) and (13.70) together give what we call the Newtonian equations. For completeness, we need to include all the constraints of the system. So, the Newtonian equations are:

$$\begin{aligned}
\dot{v}_i &= \{\{q_i, H\}_D, H\}_D \Big|_{p_i = \frac{\partial L}{\partial v_i}}, \\
\dot{q}_i &= v_i, \\
\varphi_s(q, p)\Big|_{p_i = \frac{\partial L}{\partial v_i}} &= 0, \qquad s = 1, ..., S, \quad i = 1, ..., N.
\end{aligned} \tag{13.74}$$

Notice that since the Newtonian equations are written using velocities rather than momenta, we express momenta by velocities and positions after calculating the Dirac's Brackets, using the definitions of momenta from the Legendre Transformations.

CHAPTER XIV

AN INTERPRETATION OF THE DIRAC'S BRACKETS

In this chapter, we will explore the deeper meaning of the Dirac's Bracket formula (13.32) introduced in Chapter XIII. It will allow us to understand the Dirac's Brackets better and build some intuitions.

1. An Interpretation of the Dirac's Brackets

Assume we have a space with variables (q,p) with the usual Poisson Brackets $\{F,G\}$, given by:

$$\{F,G\} = \sum_{i=1}^{N}\left(\frac{\partial F}{\partial q_i}\cdot\frac{\partial G}{\partial p_i} - \frac{\partial F}{\partial p_i}\cdot\frac{\partial G}{\partial q_i}\right). \tag{14.1}$$

Assume also that we have a minimum set of $2K$ Class II constraints, as given in Chapter XIII in formula (13.23):

$$\chi_b = \chi_b(q,p) = 0, \quad b = 1,\ldots,2K. \tag{14.2}$$

We will also assume that the constraints matrix $C^{-1} = C^{-1}(q,p)$, defined in Chapter XIII, formula (13.22):

$$C^{-1} = \begin{pmatrix} \{\chi_1,\chi_1\} & \{\chi_1,\chi_2\} & \cdots & \{\chi_1,\chi_B\} \\ \{\chi_2,\chi_1\} & \vdots & \vdots & \{\chi_2,\chi_B\} \\ \vdots & \vdots & \vdots & \vdots \\ \{\chi_B,\chi_1\} & \{\chi_B,\chi_2\} & \cdots & \{\chi_B,\chi_B\} \end{pmatrix}, \tag{14.3}$$

is invertible.

Under these conditions, the Darboux theorem, in the version presented in the Appendix D, tells us that there exists a system of variables that can replace (q,p):

$$\begin{aligned}
\phi_{Q_i} &= \phi_{Q_i}(q,p) = \phi_{Q_i}\big(\chi_b(q,p)\big), & i &= 1,\ldots,K, & b &= 1,\ldots,2K, \\
\phi_{P_i} &= \phi_{P_i}(q,p) = \phi_{P_i}\big(\chi_b(q,p)\big), & i &= 1,\ldots,K, & b &= 1,\ldots,2K, \\
Q_n &= Q_n(q,p), & n &= 1,\ldots,M = N-K, \\
P_n &= P_n(q,p), & n &= 1,\ldots,M = N-K,
\end{aligned} \tag{14.4}$$

such that:

$$\begin{aligned}
\{\phi_{Q_i},\phi_{Q_j}\} &= 0, \\
\{\phi_{P_i},\phi_{P_j}\} &= 0, \\
\{\phi_{Q_i},\phi_{P_j}\} &= \delta_{ij}, \qquad i,j = 1,\ldots,K,
\end{aligned} \tag{14.5}$$

$$\begin{aligned}
\{Q_n,Q_m\} &= 0, \\
\{P_n,P_m\} &= 0, \\
\{Q_n,P_m\} &= \delta_{nm}, \qquad n,m = 1,\ldots,M,
\end{aligned} \tag{14.6}$$

$$\begin{aligned}
\{Q_n,\phi_{Q_i}\} &= 0, \\
\{Q_n,\phi_{P_i}\} &= 0, \\
\{P_n,\phi_{Q_i}\} &= 0, \\
\{P_n,\phi_{P_i}\} &= 0. \qquad n = 1,\ldots,M, \quad i = 1,\ldots,K.
\end{aligned} \tag{14.7}$$

Then, since we have:

$$\begin{aligned}
\phi_{Q_i}(q,p) &= \phi_{Q_i}\big(\chi_b(q,p)\big), & i &= 1,\ldots,K, & b &= 1,\ldots,2K, \\
\phi_{P_i}(q,p) &= \phi_{P_i}\big(\chi_b(q,p)\big), & i &= 1,\ldots,K, & b &= 1,\ldots,2K,
\end{aligned} \tag{14.8}$$

the original constraint conditions $\chi_b(q,p) = 0, \quad b = 1,\ldots,2K,$ are equivalent to:

$$\phi_{Q_i}(q,p) = \phi_{Q_i}(0,0,...,0) = const, \quad i = 1,...,K,$$
$$\phi_{P_i}(q,p) = \phi_{P_i}(0,0,...,0) = const, \quad i = 1,...,K.$$
(14.9)

The new constraints (14.9) are equivalent to the old constraints (14.2) in the sense that they produce the same points (q,p) satisfying either set of constraints.

Let us now introduce new variables in place of $\left(\phi_{Q_i}, \phi_{P_i}\right)$, $i = 1,...,K,$ by adding constants to each variable, so that the conditions (14.9) become:

$$\phi_{Q_i}(q,p) = 0, \quad i = 1,...,K,$$
$$\phi_{P_i}(q,p) = 0, \quad i = 1,...,K.$$
(14.10)

We decided not to change the symbols for the variables, so we still use $\left(\phi_{Q_i}, \phi_{P_i}\right)$, $i = 1,...,K,$ for the modified variables that satisfy (14.10).

Then, the constraints (14.2) can be replaced by the constraints (14.10).

Adding constants to variables will not change the Poisson Brackets, since they are defined by derivatives. So, the Poisson Brackets (14.5), (14.6), and (14.7) still hold when we use the modified variables $\left(\phi_{Q_i}, \phi_{P_i}\right)$, $i = 1,...,K.$

Using the formulas (7.17) and (7.18) from Chapter VII, the regular Poisson Brackets can be written in matrix notation as:

$$\{F,G\} =$$
$$= \left(\frac{\partial F}{\partial Q_1}, \cdots, \frac{\partial F}{\partial Q_M}, \frac{\partial F}{\partial P_1}, \cdots, \frac{\partial F}{\partial P_M}, \frac{\partial F}{\partial \phi_{Q_1}}, \cdots, \frac{\partial F}{\partial \phi_{Q_K}}, \frac{\partial F}{\partial \phi_{P_1}}, \cdots, \frac{\partial F}{\partial \phi_{P_K}} \right) \times$$

$$\begin{pmatrix}
\{Q_1,Q_1\} & \cdots & \{Q_1,Q_M\} & \{Q_1,P_1\} & \cdots & \{Q_1,P_M\} & \{Q_1,\phi_{Q_1}\} & \cdots & \{Q_1,\phi_{Q_K}\} & \{Q_1,\phi_{P_1}\} & \cdots & \{Q_1,\phi_{P_K}\} \\
\vdots & \ddots & \vdots & \vdots & & \vdots & \vdots & & \vdots & \vdots & & \vdots \\
\{Q_M,Q_1\} & \cdots & \{Q_M,Q_M\} & \{Q_M,P_1\} & \cdots & \{Q_M,P_M\} & \{Q_M,\phi_{Q_1}\} & \cdots & \{Q_M,\phi_{Q_K}\} & \{Q_M,\phi_{P_1}\} & \cdots & \{Q_M,\phi_{P_K}\} \\
\{P_1,Q_1\} & \cdots & \{P_1,Q_M\} & \{P_1,P_1\} & \cdots & \{P_1,P_M\} & \{P_1,\phi_{Q_1}\} & \cdots & \{P_1,\phi_{Q_K}\} & \{P_1,\phi_{P_1}\} & \cdots & \{P_1,\phi_{P_K}\} \\
\vdots & & \vdots & \vdots & \ddots & \vdots & \vdots & & \vdots & \vdots & & \vdots \\
\{P_M,Q_1\} & \cdots & \{P_M,Q_M\} & \{P_M,P_1\} & \cdots & \{P_M,P_M\} & \{P_M,\phi_{Q_1}\} & \cdots & \{P_M,\phi_{Q_K}\} & \{P_M,\phi_{P_1}\} & \cdots & \{P_M,\phi_{P_K}\} \\
\{\phi_{Q_1},Q_1\} & \cdots & \{\phi_{Q_1},Q_M\} & \{\phi_{Q_1},P_1\} & \cdots & \{\phi_{Q_1},P_M\} & \{\phi_{Q_1},\phi_{Q_1}\} & \cdots & \{\phi_{Q_1},\phi_{Q_K}\} & \{\phi_{Q_1},\phi_{P_1}\} & \cdots & \{\phi_{Q_1},\phi_{P_K}\} \\
\vdots & & \vdots & \vdots & & \vdots & \vdots & \ddots & \vdots & \vdots & & \vdots \\
\{\phi_{Q_K},Q_1\} & \cdots & \{\phi_{Q_K},Q_M\} & \{\phi_{Q_K},P_1\} & \cdots & \{\phi_{Q_K},P_M\} & \{\phi_{Q_K},\phi_{Q_1}\} & \cdots & \{\phi_{Q_K},\phi_{Q_K}\} & \{\phi_{Q_K},\phi_{P_1}\} & \cdots & \{\phi_{Q_K},\phi_{P_K}\} \\
\{\phi_{P_1},Q_1\} & \cdots & \{\phi_{P_1},Q_M\} & \{\phi_{P_1},P_1\} & \cdots & \{\phi_{P_1},P_M\} & \{\phi_{P_1},\phi_{Q_1}\} & \cdots & \{\phi_{P_1},\phi_{Q_K}\} & \{\phi_{P_1},\phi_{P_1}\} & \cdots & \{\phi_{P_1},\phi_{P_K}\} \\
\vdots & & \vdots & \vdots & & \vdots & \vdots & & \vdots & \vdots & \ddots & \vdots \\
\{\phi_{P_K},Q_1\} & \cdots & \{\phi_{P_K},Q_M\} & \{\phi_{P_K},P_1\} & \cdots & \{\phi_{P_K},P_M\} & \{\phi_{P_K},\phi_{Q_1}\} & \cdots & \{\phi_{P_K},\phi_{Q_K}\} & \{\phi_{P_K},\phi_{P_1}\} & \cdots & \{\phi_{P_K},\phi_{P_K}\}
\end{pmatrix} \times \begin{pmatrix} \frac{\partial G}{\partial Q_1} \\ \vdots \\ \frac{\partial G}{\partial Q_M} \\ \frac{\partial G}{\partial P_1} \\ \vdots \\ \frac{\partial G}{\partial P_M} \\ \frac{\partial G}{\partial \phi_{Q_1}} \\ \vdots \\ \frac{\partial G}{\partial \phi_{Q_K}} \\ \frac{\partial G}{\partial \phi_{P_1}} \\ \vdots \\ \frac{\partial G}{\partial \phi_{P_K}} \end{pmatrix}$$

(14.11)

Using the formula (13.68) from Chapter XIII and the variables from (14.4), the Dirac Brackets can be expressed in the matrix form:

$$\{F,G\}_D =$$
$$= \left(\frac{\partial F}{\partial Q_1}, \ldots, \frac{\partial F}{\partial Q_M}, \frac{\partial F}{\partial P_1}, \ldots, \frac{\partial F}{\partial P_M}, \frac{\partial F}{\partial \phi_{Q_1}}, \ldots, \frac{\partial F}{\partial \phi_{Q_K}}, \frac{\partial F}{\partial \phi_{P_1}}, \ldots, \frac{\partial F}{\partial \phi_{P_K}} \right) \times$$

$$\begin{pmatrix} \{Q_1,Q_1\}_D & \cdots & \{Q_1,Q_M\}_D & \{Q_1,P_1\}_D & \cdots & \{Q_1,P_M\}_D & \{Q_1,\phi_{Q_1}\}_D & \cdots & \{Q_1,\phi_{Q_K}\}_D & \{Q_1,\phi_{P_1}\}_D & \cdots & \{Q_1,\phi_{P_K}\}_D \\ \vdots & \ddots & \vdots & \vdots & & \vdots & \vdots & & \vdots & \vdots & & \vdots \\ \{Q_M,Q_1\}_D & \cdots & \{Q_M,Q_M\}_D & \{Q_M,P_1\}_D & \cdots & \{Q_M,P_M\}_D & \{Q_M,\phi_{Q_1}\}_D & \cdots & \{Q_M,\phi_{Q_K}\}_D & \{Q_M,\phi_{P_1}\}_D & \cdots & \{Q_M,\phi_{P_K}\}_D \\ \{P_1,Q_1\}_D & \cdots & \{P_1,Q_M\}_D & \{P_1,P_1\}_D & \cdots & \{P_1,P_M\}_D & \{P_1,\phi_{Q_1}\}_D & \cdots & \{P_1,\phi_{Q_K}\}_D & \{P_1,\phi_{P_1}\}_D & \cdots & \{P_1,\phi_{P_K}\}_D \\ \vdots & & \vdots & \vdots & \ddots & \vdots & \vdots & & \vdots & \vdots & & \vdots \\ \{P_M,Q_1\}_D & \cdots & \{P_M,Q_M\}_D & \{P_M,P_1\}_D & \cdots & \{P_M,P_M\}_D & \{P_M,\phi_{Q_1}\}_D & \cdots & \{P_M,\phi_{Q_K}\}_D & \{P_M,\phi_{P_1}\}_D & \cdots & \{P_M,\phi_{P_K}\}_D \\ \{\phi_{Q_1},Q_1\}_D & \cdots & \{\phi_{Q_1},Q_M\}_D & \{\phi_{Q_1},P_1\}_D & \cdots & \{\phi_{Q_1},P_M\}_D & \{\phi_{Q_1},\phi_{Q_1}\}_D & \cdots & \{\phi_{Q_1},\phi_{Q_K}\}_D & \{\phi_{Q_1},\phi_{P_1}\}_D & \cdots & \{\phi_{Q_1},\phi_{P_K}\}_D \\ \vdots & & \vdots & \vdots & & \vdots & \vdots & \ddots & \vdots & \vdots & & \vdots \\ \{\phi_{Q_K},Q_1\}_D & \cdots & \{\phi_{Q_K},Q_M\}_D & \{\phi_{Q_K},P_1\}_D & \cdots & \{\phi_{Q_K},P_M\}_D & \{\phi_{Q_K},\phi_{Q_1}\}_D & \cdots & \{\phi_{Q_K},\phi_{Q_K}\}_D & \{\phi_{Q_K},\phi_{P_1}\}_D & \cdots & \{\phi_{Q_K},\phi_{P_K}\}_D \\ \{\phi_{P_1},Q_1\}_D & \cdots & \{\phi_{P_1},Q_M\}_D & \{\phi_{P_1},P_1\}_D & \cdots & \{\phi_{P_1},P_M\}_D & \{\phi_{P_1},\phi_{Q_1}\}_D & \cdots & \{\phi_{P_1},\phi_{Q_K}\}_D & \{\phi_{P_1},\phi_{P_1}\}_D & \cdots & \{\phi_{P_1},\phi_{P_K}\}_D \\ \vdots & & \vdots & \vdots & & \vdots & \vdots & & \vdots & \vdots & \ddots & \vdots \\ \{\phi_{P_K},Q_1\}_D & \cdots & \{\phi_{P_K},Q_M\}_D & \{\phi_{P_K},P_1\}_D & \cdots & \{\phi_{P_K},P_M\}_D & \{\phi_{P_K},\phi_{Q_1}\}_D & \cdots & \{\phi_{P_K},\phi_{Q_K}\}_D & \{\phi_{P_K},\phi_{P_1}\}_D & \cdots & \{\phi_{P_K},\phi_{P_K}\}_D \end{pmatrix} \times \begin{pmatrix} \frac{\partial G}{\partial Q_1} \\ \vdots \\ \frac{\partial G}{\partial Q_M} \\ \frac{\partial G}{\partial P_1} \\ \vdots \\ \frac{\partial G}{\partial P_M} \\ \frac{\partial G}{\partial \phi_{Q_1}} \\ \vdots \\ \frac{\partial G}{\partial \phi_{Q_K}} \\ \frac{\partial G}{\partial \phi_{P_1}} \\ \vdots \\ \frac{\partial G}{\partial \phi_{P_K}} \end{pmatrix}.$$

(14.12)

Writing the brackets using sums, we have, for the regular Poisson Brackets:

$$\{F,G\} = \sum_{i=1}^{M}\sum_{j=1}^{M} \frac{\partial F}{\partial Q_i} \cdot \{Q_i,Q_j\} \cdot \frac{\partial G}{\partial Q_j} + \sum_{i=1}^{M}\sum_{j=1}^{M} \frac{\partial F}{\partial Q_i} \cdot \{Q_i,P_j\} \cdot \frac{\partial G}{\partial P_j} +$$

$$+ \sum_{i=1}^{M}\sum_{j=1}^{K} \frac{\partial F}{\partial Q_i} \cdot \{Q_i,\phi_{Q_j}\} \cdot \frac{\partial G}{\partial \phi_{Q_j}} + \sum_{i=1}^{M}\sum_{j=1}^{K} \frac{\partial F}{\partial Q_i} \cdot \{Q_i,\phi_{P_j}\} \cdot \frac{\partial G}{\partial \phi_{P_j}} +$$

$$+ \sum_{i=1}^{M}\sum_{j=1}^{M} \frac{\partial F}{\partial P_i} \cdot \{P_i,Q_j\} \cdot \frac{\partial G}{\partial Q_j} + \sum_{i=1}^{M}\sum_{j=1}^{M} \frac{\partial F}{\partial P_i} \cdot \{P_i,P_j\} \cdot \frac{\partial G}{\partial P_j} +$$

(continued on the next page)

$$+\sum_{i=1}^{M}\sum_{j=1}^{K}\frac{\partial F}{\partial P_i}\cdot\{P_i,\phi_{Q_j}\}\cdot\frac{\partial G}{\partial \phi_{Q_j}} + \sum_{i=1}^{M}\sum_{j=1}^{K}\frac{\partial F}{\partial P_i}\cdot\{P_i,\phi_{P_j}\}\cdot\frac{\partial G}{\partial \phi_{P_j}} +$$

$$+\sum_{i=1}^{K}\sum_{j=1}^{M}\frac{\partial F}{\partial \phi_{Q_i}}\cdot\{\phi_{Q_i},Q_j\}\cdot\frac{\partial G}{\partial Q_j} + \sum_{i=1}^{K}\sum_{j=1}^{M}\frac{\partial F}{\partial \phi_{Q_i}}\cdot\{\phi_{Q_i},P_j\}\cdot\frac{\partial G}{\partial P_j} +$$

$$+\sum_{i=1}^{K}\sum_{j=1}^{K}\frac{\partial F}{\partial \phi_{Q_i}}\cdot\{\phi_{Q_i},\phi_{Q_j}\}\cdot\frac{\partial G}{\partial \phi_{Q_j}} + \sum_{i=1}^{K}\sum_{j=1}^{K}\frac{\partial F}{\partial \phi_{Q_i}}\cdot\{\phi_{Q_i},\phi_{P_j}\}\cdot\frac{\partial G}{\partial \phi_{P_j}} + \quad (14.13)$$

$$+\sum_{i=1}^{K}\sum_{j=1}^{M}\frac{\partial F}{\partial \phi_{P_i}}\cdot\{\phi_{P_i},Q_j\}\cdot\frac{\partial G}{\partial Q_j} + \sum_{i=1}^{K}\sum_{j=1}^{M}\frac{\partial F}{\partial \phi_{P_i}}\cdot\{\phi_{P_i},P_j\}\cdot\frac{\partial G}{\partial P_j} +$$

$$+\sum_{i=1}^{K}\sum_{j=1}^{K}\frac{\partial F}{\partial \phi_{P_i}}\cdot\{\phi_{P_i},\phi_{Q_j}\}\cdot\frac{\partial G}{\partial \phi_{Q_j}} + \sum_{i=1}^{K}\sum_{j=1}^{K}\frac{\partial F}{\partial \phi_{P_i}}\cdot\{\phi_{P_i},\phi_{P_j}\}\cdot\frac{\partial G}{\partial \phi_{P_j}}.$$

Similarly, for the Dirac Brackets, we have:

$$\{F,G\}_D = \sum_{i=1}^{M}\sum_{j=1}^{M}\frac{\partial F}{\partial Q_i}\cdot\{Q_i,Q_j\}_D\cdot\frac{\partial G}{\partial Q_j} + \sum_{i=1}^{M}\sum_{j=1}^{M}\frac{\partial F}{\partial Q_i}\cdot\{Q_i,P_j\}_D\cdot\frac{\partial G}{\partial P_j} +$$

$$+\sum_{i=1}^{M}\sum_{j=1}^{K}\frac{\partial F}{\partial Q_i}\cdot\{Q_i,\phi_{Q_j}\}_D\cdot\frac{\partial G}{\partial \phi_{Q_j}} + \sum_{i=1}^{M}\sum_{j=1}^{K}\frac{\partial F}{\partial Q_i}\cdot\{Q_i,\phi_{P_j}\}_D\cdot\frac{\partial G}{\partial \phi_{P_j}} +$$

$$+\sum_{i=1}^{M}\sum_{j=1}^{M}\frac{\partial F}{\partial P_i}\cdot\{P_i,Q_j\}_D\cdot\frac{\partial G}{\partial Q_j} + \sum_{i=1}^{M}\sum_{j=1}^{M}\frac{\partial F}{\partial P_i}\cdot\{P_i,P_j\}_D\cdot\frac{\partial G}{\partial P_j} +$$

$$+\sum_{i=1}^{M}\sum_{j=1}^{K}\frac{\partial F}{\partial P_i}\cdot\{P_i,\phi_{Q_j}\}_D\cdot\frac{\partial G}{\partial \phi_{Q_j}} + \sum_{i=1}^{M}\sum_{j=1}^{K}\frac{\partial F}{\partial P_i}\cdot\{P_i,\phi_{P_j}\}_D\cdot\frac{\partial G}{\partial \phi_{P_j}} +$$

$$+\sum_{i=1}^{K}\sum_{j=1}^{M}\frac{\partial F}{\partial \phi_{Q_i}}\cdot\{\phi_{Q_i},Q_j\}_D\cdot\frac{\partial G}{\partial Q_j} + \sum_{i=1}^{K}\sum_{j=1}^{M}\frac{\partial F}{\partial \phi_{Q_i}}\cdot\{\phi_{Q_i},P_j\}_D\cdot\frac{\partial G}{\partial P_j} +$$

$$+\sum_{i=1}^{K}\sum_{j=1}^{K}\frac{\partial F}{\partial \phi_{Q_i}}\cdot\{\phi_{Q_i},\phi_{Q_j}\}_D\cdot\frac{\partial G}{\partial \phi_{Q_j}} + \sum_{i=1}^{K}\sum_{j=1}^{K}\frac{\partial F}{\partial \phi_{Q_i}}\cdot\{\phi_{Q_i},\phi_{P_j}\}_D\cdot\frac{\partial G}{\partial \phi_{P_j}} +$$

$$+\sum_{i=1}^{K}\sum_{j=1}^{M}\frac{\partial F}{\partial \phi_{P_i}}\cdot\{\phi_{P_i},Q_j\}_D\cdot\frac{\partial G}{\partial Q_j} + \sum_{i=1}^{K}\sum_{j=1}^{M}\frac{\partial F}{\partial \phi_{P_i}}\cdot\{\phi_{P_i},P_j\}_D\cdot\frac{\partial G}{\partial P_j} +$$

$$+\sum_{i=1}^{K}\sum_{j=1}^{K}\frac{\partial F}{\partial \phi_{P_i}}\cdot\{\phi_{P_i},\phi_{Q_j}\}_D\cdot\frac{\partial G}{\partial \phi_{Q_j}} + \sum_{i=1}^{K}\sum_{j=1}^{K}\frac{\partial F}{\partial \phi_{P_i}}\cdot\{\phi_{P_i},\phi_{P_j}\}_D\cdot\frac{\partial G}{\partial \phi_{P_j}}. \quad (14.14)$$

Now, following Dirac, we define the constraint's matrix C^{-1} as:

$$C^{-1} = \begin{pmatrix} \{\phi_{Q_1}, \phi_{Q_1}\} & \cdots & \{\phi_{Q_1}, \phi_{Q_K}\} & \{\phi_{Q_1}, \phi_{P_1}\} & \cdots & \{\phi_{Q_1}, \phi_{P_K}\} \\ \vdots & \ddots & \vdots & \vdots & \ddots & \vdots \\ \{\phi_{P_K}, \phi_{Q_1}\} & \cdots & \{\phi_{Q_K}, \phi_{Q_K}\} & \{\phi_{Q_K}, \phi_{P_1}\} & \cdots & \{\phi_{Q_K}, \phi_{P_K}\} \\ \{\phi_{P_1}, \phi_{Q_1}\} & \cdots & \{\phi_{P_1}, \phi_{Q_K}\} & \{\phi_{P_1}, \phi_{P_1}\} & \cdots & \{\phi_{P_1}, \phi_{P_K}\} \\ \vdots & \ddots & \vdots & \vdots & \ddots & \vdots \\ \{\phi_{P_K}, \phi_{P_1}\} & \cdots & \{\phi_{P_K}, \phi_{Q_K}\} & \{\phi_{P_K}, \phi_{P_1}\} & \cdots & \{\phi_{P_K}, \phi_{P_K}\} \end{pmatrix}.$$

(14.15)

Placing the values from (14.5) into (14.15), we get:

$$C^{-1} = \left(\begin{array}{ccc|ccc} 0 & \cdots & 0 & 1 & 0 & \cdots & 0 \\ \vdots & \ddots & \vdots & 0 & \ddots & & \vdots \\ \vdots & & \ddots & \vdots & \vdots & & \ddots & 0 \\ 0 & \cdots & \cdots & 0 & 0 & \cdots & \cdots & 1 \\ \hline -1 & 0 & \cdots & 0 & 0 & \cdots & \cdots & 0 \\ 0 & \ddots & & \vdots & \vdots & \ddots & & \vdots \\ \vdots & & \ddots & 0 & \vdots & & \ddots & \vdots \\ 0 & \cdots & \cdots & -1 & 0 & \cdots & \cdots & 0 \end{array}\right).$$

(14.16)

The lines inside the matrix C^{-1} do not have any mathematical meaning. They are there only to help to better see the structure of the matrix. We can see that the matrix C^{-1} is made of four square matrices. Two of them have all entries equal to zero; the other two have non-zero entries only on their diagonals.

Because of this simple structure, it is easy to guess the matrix that is an inverse to the matrix C^{-1}, as:

$$C = \left(\begin{array}{ccc|ccc} 0 & \cdots & \cdots & 0 & -1 & 0 & \cdots & 0 \\ \vdots & \ddots & & \vdots & 0 & \ddots & & \vdots \\ \vdots & & \ddots & \vdots & \vdots & & \ddots & 0 \\ 0 & \cdots & \cdots & 0 & 0 & \cdots & \cdots & -1 \\ \hline 1 & 0 & \cdots & 0 & 0 & \cdots & \cdots & 0 \\ 0 & \ddots & & \vdots & \vdots & \ddots & & \vdots \\ \vdots & & \ddots & 0 & \vdots & & \ddots & \vdots \\ 0 & \cdots & \cdots & 1 & 0 & \cdots & \cdots & 0 \end{array}\right).$$

(14.17)

By the direct multiplication, we can easily check that multiplying CC^{-1}, we get the identity matrix, so the matrix C is indeed the inverse of the matrix C^{-1}.

Placing C into the definition of the Dirac's Brackets (13.24), we get:

$$\{F,G\}_D = \{F,G\} - \left(\{F,\phi_{Q_1}\},...,\{F,\phi_{Q_K}\},\{F,\phi_{P_1}\},...,\{F,\phi_{P_K}\}\right) \times$$

$$\times \begin{pmatrix} 0 & \cdots & \cdots & 0 & -1 & 0 & \cdots & 0 \\ \vdots & \ddots & & \vdots & 0 & \ddots & & \vdots \\ \vdots & & \ddots & \vdots & \vdots & & \ddots & 0 \\ 0 & \cdots & \cdots & 0 & 0 & \cdots & \cdots & -1 \\ 1 & 0 & \cdots & 0 & 0 & \cdots & \cdots & 0 \\ 0 & \ddots & & \vdots & \vdots & \ddots & & \vdots \\ \vdots & & \ddots & 0 & \vdots & & \ddots & \vdots \\ 0 & \cdots & \cdots & 1 & 0 & \cdots & \cdots & 0 \end{pmatrix} \begin{pmatrix} \{\phi_{Q_1},G\} \\ \vdots \\ \{\phi_{Q_K},G\} \\ \{\phi_{P_1},G\} \\ \vdots \\ \{\phi_{P_K},G\} \end{pmatrix}. \qquad (14.18)$$

Multiplying the vector on the very right by the matrix C, we get:

$$\{F,G\}_D = \{F,G\} +$$

$$-\left(\{F,\phi_{Q_1}\},...,\{F,\phi_{Q_K}\},\{F,\phi_{P_1}\},...,\{F,\phi_{P_K}\}\right) \begin{pmatrix} -\{\phi_{P_1},G\} \\ \vdots \\ -\{\phi_{P_K},G\} \\ \{\phi_{Q_1},G\} \\ \vdots \\ \{\phi_{Q_K},G\} \end{pmatrix} \qquad (14.19)$$

Multiplying the two vectors, we get:

$$\{F,G\}_D = \{F,G\} + \sum_{i=1}^{K} \{F,\phi_{Q_i}\}\{\phi_{P_i},G\} - \sum_{i=1}^{K} \{F,\phi_{P_i}\}\{\phi_{Q_i},G\}. \qquad (14.20)$$

Now we can use (14.20) to calculate the Dirac Brackets of the variables. We also use (14.5), (14.6) and (14.7):

$$\{\phi_{Q_n}, \phi_{Q_m}\}_D = \{\phi_{Q_n}, \phi_{Q_m}\} + \sum_{i=1}^{K}\{\phi_{Q_n}, \phi_{Q_i}\}\{\phi_{P_i}, \phi_{Q_m}\} - \sum_{i=1}^{K}\{\phi_{Q_n}, \phi_{P_i}\}\{\phi_{Q_i}, \phi_{Q_m}\} =$$

$$= 0 + \sum_{i=1}^{K} 0 \cdot (-\delta_{im}) - \sum_{i=1}^{K} \delta_{ni} \cdot 0 = 0,$$

$$\{\phi_{P_n}, \phi_{P_m}\}_D = \{\phi_{P_n}, \phi_{P_m}\} + \sum_{i=1}^{K}\{\phi_{P_n}, \phi_{Q_i}\}\{\phi_{P_i}, \phi_{P_m}\} - \sum_{i=1}^{K}\{\phi_{P_n}, \phi_{P_i}\}\{\phi_{Q_i}, \phi_{P_m}\} =$$

$$= 0 + \sum_{i=1}^{K} (-\delta_{ni}) \cdot 0 - \sum_{i=1}^{K} 0 \cdot \delta_{im} = 0,$$

$$\{\phi_{Q_n}, \phi_{P_m}\}_D = \{\phi_{Q_n}, \phi_{P_m}\} + \sum_{i=1}^{K}\{\phi_{Q_n}, \phi_{Q_i}\}\{\phi_{P_i}, \phi_{P_m}\} - \sum_{i=1}^{K}\{\phi_{Q_n}, \phi_{P_i}\}\{\phi_{Q_i}, \phi_{P_m}\} =$$

$$= \delta_{nm} + \sum_{i=1}^{K} 0 \cdot 0 - \sum_{i=1}^{K} \delta_{ni} \cdot \delta_{im} = \delta_{nm} - \delta_{nm} = 0,$$

$$\{Q_n, Q_m\}_D = \{Q_n, Q_m\} + \sum_{i=1}^{K}\{Q_n, \phi_{Q_i}\}\{\phi_{P_i}, Q_m\} - \sum_{i=1}^{K}\{Q_n, \phi_{P_i}\}\{\phi_{Q_i}, Q_m\} =$$

$$= 0 + \sum_{i=1}^{K} 0 \cdot 0 - \sum_{i=1}^{K} 0 \cdot 0 = 0,$$

$$\{P_n, P_m\}_D = \{P_n, P_m\} + \sum_{i=1}^{K}\{P_n, \phi_{Q_i}\}\{\phi_{P_i}, P_m\} - \sum_{i=1}^{K}\{P_n, \phi_{P_i}\}\{\phi_{Q_i}, P_m\} =$$

$$= 0 + \sum_{i=1}^{K} 0 \cdot 0 - \sum_{i=1}^{K} 0 \cdot 0 = 0,$$

$$\{Q_n, P_m\}_D = \{Q_n, P_m\} + \sum_{i=1}^{K}\{Q_n, \phi_{Q_i}\}\{\phi_{P_i}, P_m\} - \sum_{i=1}^{K}\{Q_n, \phi_{P_i}\}\{\phi_{Q_i}, P_m\} =$$

$$= \delta_{nm} + \sum_{i=1}^{K} 0 \cdot 0 - \sum_{i=1}^{K} 0 \cdot 0 = \delta_{nm},$$

$$\{Q_n, \phi_{Q_m}\}_D = \{Q_n, \phi_{Q_m}\} + \sum_{i=1}^{K}\{Q_n, \phi_{Q_i}\}\{\phi_{P_i}, \phi_{Q_m}\} - \sum_{i=1}^{K}\{Q_n, \phi_{P_i}\}\{\phi_{Q_i}, \phi_{Q_m}\} =$$

$$= 0 + \sum_{i=1}^{K} 0 \cdot (-\delta_{im}) - \sum_{i=1}^{K} 0 \cdot 0 = 0,$$

$$\{Q_n, \phi_{P_m}\}_D = \{Q_n, \phi_{P_m}\} + \sum_{i=1}^{K}\{Q_n, \phi_{Q_i}\}\{\phi_{P_i}, \phi_{P_m}\} - \sum_{i=1}^{K}\{Q_n, \phi_{P_i}\}\{\phi_{Q_i}, \phi_{P_m}\} =$$

$$= 0 + \sum_{i=1}^{K} 0 \cdot 0 - \sum_{i=1}^{K} 0 \cdot \delta_{im} = 0,$$

$$\{P_n, \phi_{Q_m}\}_D = \{P_n, \phi_{Q_m}\} + \sum_{i=1}^{K}\{P_n, \phi_{Q_i}\}\{\phi_{P_i}, \phi_{Q_m}\} - \sum_{i=1}^{K}\{P_n, \phi_{P_i}\}\{\phi_{Q_i}, \phi_{P_m}\} =$$

$$= 0 + \sum_{i=1}^{K} 0 \cdot (-\delta_{im}) - \sum_{i=1}^{K} 0 \cdot 0 = 0,$$

$$\{P_n, \phi_{P_m}\}_D = \{P_n, \phi_{P_m}\} + \sum_{i=1}^{K}\{P_n, \phi_{Q_i}\}\{\phi_{P_i}, \phi_{P_m}\} - \sum_{i=1}^{K}\{P_n, \phi_{P_i}\}\{\phi_{P_i}, \phi_{P_m}\} =$$

$$= 0 + \sum_{i=1}^{K} 0 \cdot 0 - \sum_{i=1}^{K} 0 \cdot \delta_{im} = 0.$$

Now, we place the above in (14.14), getting:

$$\{F,G\}_D = \sum_{i=1}^{M}\sum_{j=1}^{M}\frac{\partial F}{\partial Q_i}\cdot 0 \cdot \frac{\partial G}{\partial Q_j} + \sum_{i=1}^{M}\sum_{j=1}^{M}\frac{\partial F}{\partial Q_i}\cdot \delta_{ij}\cdot \frac{\partial G}{\partial P_j} +$$

$$+\sum_{i=1}^{M}\sum_{j=1}^{K}\frac{\partial F}{\partial Q_i}\cdot 0 \cdot \frac{\partial G}{\partial \phi_{Q_j}} + \sum_{i=1}^{M}\sum_{j=1}^{K}\frac{\partial F}{\partial Q_i}\cdot 0 \cdot \frac{\partial G}{\partial \phi_{P_j}} +$$

$$+\sum_{i=1}^{M}\sum_{j=1}^{M}\frac{\partial F}{\partial P_i}\cdot (-\delta_{ij}) \cdot \frac{\partial G}{\partial Q_j} + \sum_{i=1}^{M}\sum_{j=1}^{M}\frac{\partial F}{\partial P_i}\cdot 0 \cdot \frac{\partial G}{\partial P_j} +$$

$$+\sum_{i=1}^{M}\sum_{j=1}^{K}\frac{\partial F}{\partial P_i}\cdot 0 \cdot \frac{\partial G}{\partial \phi_{Q_j}} + \sum_{i=1}^{M}\sum_{j=1}^{K}\frac{\partial F}{\partial P_i}\cdot 0 \cdot \frac{\partial G}{\partial \phi_{P_j}} +$$

$$+\sum_{i=1}^{K}\sum_{j=1}^{M}\frac{\partial F}{\partial \phi_{Q_i}}\cdot 0 \cdot \frac{\partial G}{\partial Q_j} + \sum_{i=1}^{K}\sum_{j=1}^{M}\frac{\partial F}{\partial \phi_{Q_i}}\cdot 0 \cdot \frac{\partial G}{\partial P_j} +$$

$$+\sum_{i=1}^{K}\sum_{j=1}^{K}\frac{\partial F}{\partial \phi_{Q_i}}\cdot 0 \cdot \frac{\partial G}{\partial \phi_{Q_j}} + \sum_{i=1}^{K}\sum_{j=1}^{K}\frac{\partial F}{\partial \phi_{Q_i}}\cdot 0 \cdot \frac{\partial G}{\partial \phi_{P_j}} +$$

$$+\sum_{i=1}^{K}\sum_{j=1}^{M}\frac{\partial F}{\partial \phi_{P_i}}\cdot 0 \cdot \frac{\partial G}{\partial Q_j} + \sum_{i=1}^{K}\sum_{j=1}^{M}\frac{\partial F}{\partial \phi_{P_i}}\cdot 0 \cdot \frac{\partial G}{\partial P_j} +$$

$$+\sum_{i=1}^{K}\sum_{j=1}^{K}\frac{\partial F}{\partial \phi_{P_i}}\cdot 0 \cdot \frac{\partial G}{\partial \phi_{Q_j}} + \sum_{i=1}^{K}\sum_{j=1}^{K}\frac{\partial F}{\partial \phi_{P_i}}\cdot 0 \cdot \frac{\partial G}{\partial \phi_{P_j}}.$$

So:

$$\{F,G\}_D = \sum_{i=1}^{M}\frac{\partial F}{\partial Q_i}\cdot \frac{\partial G}{\partial P_i} - \sum_{i=1}^{M}\frac{\partial F}{\partial P_i}\cdot \frac{\partial G}{\partial Q_i}. \tag{14.21}$$

Notice that using (14.21) for (Q_m, P_m) we get:

$$\{Q_n, Q_m\}_D = 0,$$
$$\{P_n, P_m\}_D = 0, \qquad (14.22)$$
$$\{Q_n, P_m\}_D = \delta_{nm}, \qquad n, m = 1, ..., M.$$

Comparing (14.22) with (14.6) tells us that for functions depending on (Q_m, P_m) only, the Dirac's Brackets are identical to the original Poisson Brackets.

So, one possible interpretation of the definition of the Dirac's Brackets is that their use is equivalent to switching from the original canonical variables (q_i, p_i), $i = 1, ..., N$, to the canonical variables of Class I, (Q_m, P_m), $m = 1, ..., M$, and canonical variables of Class II, (ϕ_{Q_k}, ϕ_{P_k}), $k = M+1, ..., N$, and then leaving only the functions of the variables (Q_m, P_m) in the formalism, while still using the regular Poisson Brackets in the form (14.21).

Notice that using only the functions of (Q_m, P_m) in the formalism is justified by the fact that the variables (ϕ_{Q_k}, ϕ_{P_k}) have time derivatives equal to zero, so these variables can be replaced by constants in any function, without changing the time dependence of such function.

An important feature of the formula (14.21) is that it is structurally identical to the definition of the Poisson Bracket. So, all the basic properties of the Dirac's Brackets, given in Chapter XIII as formulas from (13.25) to (13.34), do not require a separate proof because the proof is going to be identical to the proof shown earlier for the Poisson Brackets.

CHAPTER XV

COMMENTS ON DIRAC'S BRACKETS

We want to make some comments here. They are of a general nature. Specific examples of using some of the concepts described below can be found in Volumes II and III of this book.

1. Calculating the Dirac's Brackets One Pair or Few Pairs of Constraints at a Time

When we look at the formula for Dirac's Brackets, which is (13.23) in Chapter XIII, we notice that we need to calculate an inverse of a matrix. The entries of this matrix are not numbers, they are formulas. Calculating inverses of large matrices, especially if their entries are formulas, is not easy, and the number of operations needed increases proportionally to the factorial of the matrix size. So, even computerized calculations may quickly become too large to handle.

The alternative to using the formula (13.23) for all constraints of Class II of the system is to do it one pair of constraints at a time. To start, we need to choose such two constraints χ_1, χ_2 that have non-zero Poisson Brackets. Calculating the inverse of a 2x2 antisymmetric matrix is trivial. So, calculating the Dirac's Brackets for just these two constraints will be simple. We can call the result the Dirac's Brackets of the first generation, $\{F,G\}_{D1}$.

Then, we repeat the process with the next pair of constraints. We will have to use the Dirac's Brackets that we got in the first step in the role of the Poisson Brackets in the definition of the Dirac's Brackets of the second generation. Also, the pair of the constraints we use, χ_3, χ_4 must satisfy $\{\chi_3, \chi_4\}_{D1} \neq 0$. Then we define the Dirac's Brackets of the second generation as:

$$\{F,G\}_{D2} = \{F,G\}_{D1} - \sum_{b=3}^{4}\sum_{c=3}^{4} \{F,\chi_b\}_{D1} \cdot [C_{bc}]_{D1} \cdot \{\chi_c,G\}_{D1}, \qquad (15.1)$$

where $[C_{bc}]_{D1}$ is the matrix obtained from the constraints χ_3, χ_4, using the Dirac's Brackets of the first generation.

The process is then repeated until all constraints are included.

Will this process arrive at the same the same Dirac's Brackets as doing it in one step? The answer is yes. In Appendix D, the reader can find the construction of the Darboux variables. It should be visible that we could adjust the pairs in the construction of the Darboux variables to the pairs of constraints used in the process above. Then, we could use the same Darboux variables as constraints to calculate the Dirac's Brackets in one large step, or as a combination of smaller steps. In both cases the ending result will be the removal of all these variables from the formula for the Dirac's Brackets. Compare the formula (14.21) in Chapter XV.

2. Using Only Some Constraints to Define the Dirac's Brackets

In practice, we do not have to use all the constraints of Class II to define Dirac's Brackets. It may be convenient to only use some of them and then stop there.

It may also be convenient to use some constraints, make conclusions about the system, and then finish the process using the rest of the constraints.

3. Imposing Additional Constraints on The System

In the text above, we were only considering constraints that originated from the Euler-Lagrange equations of the system. However, we do not need to restrict ourselves only to these constraints. We can impose other, arbitrary, constraints on the system.

Actually, what we impose may be not only constraints in the sense we had earlier, but also families of constraints where, instead of zero on one side, we have a constant there. For example

$$\varphi = q_1^2 + q_2^2 = r^2, \tag{15.2}$$

can be treated the same way as other constraints were.

When we do that, we need to go through the process of completing the system of Euler-Lagrange equations for these imposed constraints as well. In other words, the time derivatives of each constraint must be equal to zero, exactly as for the original constraints, and this may produce more constraints. Time derivatives of these resulting constraints must also be zero, and so on. Then all the constraints of the system are again divided into Class I and II, and the process is analogous.

Notice that this process may result in contradiction, namely a constraint like $1=0$. This would mean that the system of such equations has no solutions and should be discarded. As we had shown earlier, something similar may happen for constraints that come directly from the Lagrangian as well.

Also, it is important that we understand that imposing constraints on a system in the way described above is very different from imposing real-life mechanical constraints on a system. The difference is that imposing constraints by the "mathematical" process above does not change the original differential equations of motion that are present in the system. Instead, the process of completing the system requires adding more constraints, so that the entire system becomes contradiction-free (or collapses completely). In a sense, it is a process of generating the set of all initial conditions that need to be imposed on a system, which then will hold as the system is evolving with time.

This is very different from imposing real-life constraints on real-life mechanical systems. Real-life constraints create extra physical forces in a mechanical system, and these forces show up as modifications of the original differential equations of motion. This is very different from the situation described above, where the differential equations are not modified.

We may also notice that families of constraints, as in (15.2), may be used to fix the "gauge" in non-regular (gauge) systems. If chosen correctly, they will fix the "missing" velocities in the non-regular systems. As we mentioned before, a more detailed analysis of gauge systems is outside of the scope of this book.

4. Imposing Constants of Motion as Constraints

In point 3 above, we mentioned the possibility of imposing an extra family of constraints on the system. Notice that such constraints' family may be obtained from an existing constant of motion, by imposing a condition that this constant of motion is equal to a specific number. Then, the time derivative of such constraint will be automatically zero, and no more constraints will be produced by the completion process.

If we then find another constant of motion that has non-zero Poisson Bracket (or Dirac's Bracket) with the first one, then we can use both for defining modified Dirac's Brackets. This would, in a sense, remove both constants of motion from the dynamics of the system.

Notice that, if we try to use the Hamiltonian as one of the constants above, then there exists no constant of motion that would have non-zero Poisson Bracket (or Dirac's Bracket) with the Hamiltonian, since that Poisson Bracket (or Dirac's Bracket) is equal to the time derivative of the constant of motion, so it would have to be zero. So, the Hamiltonian cannot be used to define the modified Dirac's Brackets.

How many constants of motion can be used to define such new Dirac Brackets? Actually, we can involve all the constants of motion, except the Hamiltonian.

If we look at the proof of the Darboux Theorem presented in Appendix D, we may observe that the process of finding the Darboux variables can start with choosing the Hamiltonian as P_1. Then, we get the canonical variables $(Q_1, P_1, Q_2, P_2, ..., Q_N, P_N)$, where $P_1 = H$. Through the Darboux construction, we have

$$\begin{aligned}\dot{Q}_i &= \{Q_i, H\} = \{Q_i, P_1\} = 0, \quad i = 2, ..., N, \\ \dot{P}_j &= \{P_j, H\} = \{Q_j, P_1\} = 0. \quad j = 1, ..., N.\end{aligned} \quad (15.3)$$

So, all these variables except Q_1 are the constants of motion. They come in pairs (Q_i, P_i), $i = 2, ..., N$, with the $\{Q_i, P_i\} = 1$, $i = 2, ..., N$.

It can be easily checked that if we introduce constraints based on these pairs by making each one equal to a number and then calculate Dirac's Brackets using these constraints, we get:

$$\{Q_1, P_1\}_D = 1,$$

(continued on the next page)

$$\{Q_1, P_k\}_D = 0, \quad k = 2,\ldots,N,$$
$$\{Q_k, P_1\}_D = 0, \quad k = 2,\ldots,N,$$
$$\{Q_i, P_j\}_D = 0, \quad i,j = 2,\ldots,N, \quad (15.4)$$
$$\{Q_n, Q_m\}_D = 0, \quad m,n = 1,\ldots,N,$$
$$\{P_n, P_m\}_D = 0, \quad m,n = 1,\ldots,N.$$

So, the dynamical brackets of this system degenerate to just two non-trivial dimensions.

Technically speaking, it gives correct equations of motion. It seems that the dynamical brackets system when using this construction is trivialized too much though. However, it also seems to be an arbitrary decision how we do it. Whether it makes sense for physical systems cannot be decided solely based on mathematics of the problem.

Notice that the variable Q_1 in the above can be interpreted as a "clock" in the system. This is because its time derivative is always equal to 1, so it can be used to measure the time parameter. Since mathematically we have quite a lot of freedom to choose the Q_1, we have many possible "clocks".

Notice that since $\{Q_1, P_1\} = \{Q_1, H\} = 1$, after the quantization, we get $\{\hat{Q}_1, \hat{H}\} = i\hbar \hat{1}$. This is related to the fact that the Heisenberg Uncertainty Principle will tell us that the precise measurement of time and energy (Hamiltonian) is impossible. Notice that time is an organizing parameter that can be known precisely in classical theoretical considerations, but it is also a variable that in the real (quantum) world must be experimentally measured using $\hat{Q}_1$. In the quantum setting, it cannot be measured with infinite precision.

5. A Comment about Volumes II and III

In Volumes II and III of this text, we will consider examples and applications. Some of the issues mentioned above will be addressed there in more detail.

APPENDIX A

THE FLOW BOX THEOREM

The flow box theorem is also known under other names (the straightening out theorem, the domain-straightening theorem, straightening out of a vector field theorem, or the box theorem). The theorem may be expressed in multiple ways. We will present it here in a way we see best fitting our use in this text.

The flow box theorem assumes that we have a system of ordinary differential equations:

$$\begin{aligned}\dot{x}_1 &= f_1(x_1, x_2, ..., x_n), \\ \dot{x}_2 &= f_2(x_1, x_2, ..., x_n), \\ &\vdots \\ \dot{x}_n &= f_n(x_1, x_2, ..., x_n),\end{aligned} \qquad (A.1)$$

with the independent variable denoted as t, and other variables denoted as $(x_1, x_2, ..., x_n)$. On the right side of (A.1), we have smooth enough functions $(f_1, f_2, ..., f_n)$, and the dots above the variables denote the derivatives with respect to t.

Then, the flow box theorem states that, for any given point $(x_{0,1}, x_{0,2}, ..., x_{0,n})$, such that not all functions $(f_1, f_2, ..., f_n)$ are equal to zero at that point, there exists such neighborhood of the point and such a change of variables:

$$\begin{aligned}y_1 &= y_1(x_1, x_2, ..., x_n), \\ y_2 &= y_2(x_1, x_2, ..., x_n), \\ &\vdots \\ y_n &= y_n(x_1, x_2, ..., x_n),\end{aligned} \qquad (A.2)$$

in the neighborhood, that the equations (A.1) written in the new variables in this neighborhood are:

$$\dot{y}_1 = 1,$$
$$\dot{y}_2 = 0,$$
$$\vdots \qquad\qquad\qquad\qquad\qquad\qquad\qquad\qquad\qquad\qquad\qquad\qquad (A.3)$$
$$\dot{y}_n = 0.$$

Sketch of the proof:

We will describe a sequence of changes of variables that, when combined, give the change of variables that satisfy the theorem.

The first change of the variables is done by adding a possibly different constant to each existing variable to get a new one, in such a way that the point $(x_{0,1}, x_{0,2}, ..., x_{0,n})$ described in the theorem in the new variables become $(0,0,...,0)$. It is not convenient to change to new symbols describing the new variables. So, we stay with the old variable symbols $(x_1, x_2, ..., x_n)$, but now we assume that the point described in the theorem has the coordinates $(0,0,...,0)$. Also, we will keep the same form of the equations (A.1). Obviously, after the change of variables, the specific functions in (A.1) will change their form, but we decide to still use the same symbols for the modified function. So we still use (A.1) as our equations.

The second change of the variables is a "rotation" that has the point $(0,0,...,0)$ at its center. Since the variables $(x_1, x_2, ..., x_n)$ do not have to be Cartesian, this is not a real rotation. Still, we use an invertible matrix M_{ij}, $i,j = 1,...,n,$ made of constant numbers, not functions of $(x_1, x_2, ..., x_n)$, to change the variables in a neighborhood of the point $(0,0,...,0)$. Specifically, we define the new variables as:

$$\begin{pmatrix} z_1 \\ z_2 \\ \vdots \\ z_n \end{pmatrix} = \begin{pmatrix} M_{11} & M_{12} & \cdots & M_{1n} \\ M_{21} & M_{22} & \cdots & M_{2n} \\ \vdots & \vdots & & \vdots \\ M_{n1} & M_{n2} & \cdots & M_{nn} \end{pmatrix} \begin{pmatrix} x_1 \\ x_2 \\ \vdots \\ x_n \end{pmatrix}. \qquad (A.4)$$

Since it is a rotation about the origin, then the point with coordinates $(0,0,...,0)$ will still be $(0,0,...,0)$ in the new variables as well.

Assuming that variables $(x_1, x_2, ..., x_n)$, when considered as solutions of the equations (A.1), are functions of time t, then so are $(z_1, z_2, ..., z_n)$ via the equation (A.4). Taking the time derivative of both sides of (A.4), with the matrix M made of constant numbers, we get:

$$\begin{pmatrix} \dot{z}_1 \\ \dot{z}_2 \\ \vdots \\ \dot{z}_n \end{pmatrix} = \begin{pmatrix} M_{11} & M_{12} & \cdots & M_{1n} \\ M_{21} & M_{22} & \cdots & M_{2n} \\ \vdots & \vdots & & \vdots \\ M_{n1} & M_{n2} & \cdots & M_{nn} \end{pmatrix} \begin{pmatrix} \dot{x}_1 \\ \dot{x}_2 \\ \vdots \\ \dot{x}_n \end{pmatrix}. \qquad (A.5)$$

Placing time derivatives of $(x_1, x_2, ..., x_n)$ from (A.1) in (A.5) we get:

$$\begin{pmatrix} \dot{z}_1 \\ \dot{z}_2 \\ \vdots \\ \dot{z}_n \end{pmatrix} = \begin{pmatrix} M_{11} & M_{12} & \cdots & M_{1n} \\ M_{21} & M_{22} & \cdots & M_{2n} \\ \vdots & \vdots & & \vdots \\ M_{n1} & M_{n2} & \cdots & M_{nn} \end{pmatrix} \begin{pmatrix} f_1(x_1, x_2, ..., x_n) \\ f_2(x_1, x_2, ..., x_n) \\ \vdots \\ f_n(x_1, x_2, ..., x_n) \end{pmatrix}. \qquad (A.6)$$

(A.6) represents the equations (A.1), with the left side expressed in variables $(z_1, z_2, ..., z_n)$, and the right side in variables $(x_1, x_2, ..., x_n)$.

The assumption was that at the point $(0,0,...,0)$ not all functions on the right side of (A.6) were equal to zero. So, we obtain from the elementary algebra that the invertible matrix M in (A.6) can be chosen is such a way that we get:

$$\begin{pmatrix} 1 \\ 0 \\ \vdots \\ 0 \end{pmatrix} = \begin{pmatrix} M_{11} & M_{12} & \cdots & M_{1n} \\ M_{21} & M_{22} & \cdots & M_{2n} \\ \vdots & \vdots & & \vdots \\ M_{n1} & M_{n2} & \cdots & M_{nn} \end{pmatrix} \begin{pmatrix} f_1(0,0,...,0) \\ f_2(0,0,...,0) \\ \vdots \\ f_n(0,0,...,0) \end{pmatrix}. \tag{A.7}$$

Actually, there are many such matrices. We can arbitrarily choose one of them and define the variables $(z_1, z_2, ..., z_n)$ using this specific one.

Then we can define functions $(g_1, g_2, ..., g_n)$ of $(z_1, z_2, ..., z_n)$ as:

$$\begin{pmatrix} g_1(z_1,z_2,...,z_n) \\ g_2(z_1,z_2,...,z_n) \\ \vdots \\ g_n(z_1,z_2,...,z_n) \end{pmatrix} = \begin{pmatrix} M_{11} & M_{12} & \cdots & M_{1n} \\ M_{21} & M_{22} & \cdots & M_{2n} \\ \vdots & \vdots & & \vdots \\ M_{n1} & M_{n2} & \cdots & M_{nn} \end{pmatrix} \begin{pmatrix} f_1(x_1,x_2,...,x_n) \\ f_2(x_1,x_2,...,x_n) \\ \vdots \\ f_n(x_1,x_2,...,x_n) \end{pmatrix}, \tag{A.8}$$

where $(z_1, z_2, ..., z_n)$ on the left side are obtained from $(x_1, x_2, ..., x_n)$ on the right side via equations (A.4).

Then using (A.8) in (A.6), we obtain:

$$\begin{aligned} \dot{z}_1 &= g_1(z_1, z_2, ..., z_n), \\ \dot{z}_2 &= g_2(z_1, z_2, ..., z_n), \\ &\vdots \\ \dot{z}_n &= g_n(z_1, z_2, ..., z_n). \end{aligned} \tag{A.9}$$

The equations (A.9) are the equations (A.1) expressed in the variables $(z_1, z_2, ..., z_n)$.

Placing the point $(0, 0, ..., 0)$ on both sides of (A.8), we get:

$$\begin{pmatrix} g_1(0,0,...,0) \\ g_2(0,0,...,0) \\ \vdots \\ g_n(0,0,...,0) \end{pmatrix} = \begin{pmatrix} M_{11} & M_{12} & \cdots & M_{1n} \\ M_{21} & M_{22} & \cdots & M_{2n} \\ \vdots & \vdots & & \vdots \\ M_{n1} & M_{n2} & \cdots & M_{nn} \end{pmatrix} \begin{pmatrix} f_1(0,0,...,0) \\ f_2(0,0,...,0) \\ \vdots \\ f_n(0,0,...,0) \end{pmatrix}. \tag{A.10}$$

Comparing (A.10) with (A.7), we get:

$$\begin{aligned} g_1(0,0,...,0) &= 1, \\ g_2(0,0,...,0) &= 0, \\ &\vdots \\ g_n(0,0,...,0) &= 0. \end{aligned} \tag{A.11}$$

Let us now look at the solutions of the equations (A.9). The Cauchy's theorem tells us that for any given initial conditions $(z_{0,1}, z_{0,2}, ..., z_{0,n})$, we have the unique solution Ψ in the form:

$$\begin{aligned} z_1 &= \Psi_1\big(t;(z_{0,1}, z_{0,2}, ..., z_{0,n})\big), \\ z_2 &= \Psi_2\big(t;(z_{0,1}, z_{0,2}, ..., z_{0,n})\big), \\ &\vdots \\ z_n &= \Psi_n\big(t;(z_{0,1}, z_{0,2}, ..., z_{0,n})\big). \end{aligned} \tag{A.12}$$

As we can see, solutions are functions of the initial conditions $(z_{0,1}, z_{0,2}, ..., z_{0,n})$ and the time parameter t. Now, let us restrict the solutions (A.12) to the initial conditions in the form $(0, z_{0,2}, ..., z_{0,n})$. We get:

$$\begin{aligned} z_1 &= \Psi_1\big(t;(0, z_{0,2}, ..., z_{0,n})\big), \\ z_2 &= \Psi_2\big(t;(0, z_{0,2}, ..., z_{0,n})\big), \\ &\vdots \\ z_n &= \Psi_n\big(t;(0, z_{0,2}, ..., z_{0,n})\big). \end{aligned} \tag{A.13}$$

Notice that, from the definition of the initial conditions when placing $t = 0$ into (A.13), we get:

$$0 = \Psi_1\left(0;\left(0, z_{0,2}, ..., z_{0,n}\right)\right),$$
$$z_{0,2} = \Psi_2\left(0;\left(0, z_{0,2}, ..., z_{0,n}\right)\right),$$
$$\vdots \qquad (A.14)$$
$$z_{0,n} = \Psi_n\left(0;\left(0, z_{0,2}, ..., z_{0,n}\right)\right).$$

Because of the Cauchy theorem, for each given point $(z_1, z_2, ..., z_n)$ in a neighborhood of the point $(0, 0, ..., 0)$, there is only one solution in the form (A.13) that passes through $(z_1, z_2, ..., z_n)$. Namely, at the point $(z_1, z_2, ..., z_n)$ we can take a solution in the form (A.12) using $(z_1, z_2, ..., z_n)$ as the initial condition. Then we can look for such $-t$ closest to zero that we get a result with the first variable equal to zero and being in the neighborhood of the point $(0, 0, ..., 0)$ chosen earlier. If this is not happening, it means that we have chosen a neighborhood that is too large, and we need to choose a smaller one. In the smaller neighborhood, using (A.12) we have:

$$0 = \Psi_1\left(-t;\left(z_1, z_2, ..., z_n\right)\right),$$
$$z_{0,2} = \Psi_2\left(-t;\left(z_1, z_2, ..., z_n\right)\right),$$
$$\vdots \qquad (A.15)$$
$$z_{0,n} = \Psi_n\left(-t;\left(z_1, z_2, ..., z_n\right)\right).$$

Reversing the time flow in (A.16), from the general properties of the solutions of the ordinary differential equations, we get:

$$z_1 = \Psi_1\left(t;\left(0, z_{0,2}, ..., z_{0,n}\right)\right),$$
$$z_2 = \Psi_2\left(t;\left(0, z_{0,2}, ..., z_{0,n}\right)\right),$$
$$\vdots \qquad (A.16)$$
$$z_n = \Psi_n\left(t;\left(0, z_{0,2}, ..., z_{0,n}\right)\right).$$

The relations (A.15) and (A.16) show one-to-one correspondence between the variables $(z_1, z_2, ..., z_n)$ and the variables $(t; (0, z_{0,2}, ..., z_{0,n}))$.

So, we can introduce new variables $(y_1, y_2, ..., y_n)$ as:

$$y_1 = t,$$
$$y_2 = z_{0,2},$$
$$\vdots \qquad\qquad\qquad\qquad\qquad\qquad\qquad\qquad\qquad\qquad\qquad\text{(A.17)}$$
$$y_n = z_{0,n}.$$

Then from (A.16), we have the change of variables from $(y_1, y_2, ..., y_n)$ to $(z_1, z_2, ..., z_n)$ given as:

$$z_1 = \Psi_1(y_1; (0, y_2, ..., y_n)),$$
$$z_2 = \Psi_2(y_1; (0, y_2, ..., y_n)),$$
$$\vdots \qquad\qquad\qquad\qquad\qquad\qquad\qquad\qquad\qquad\qquad\text{(A.18)}$$
$$z_n = \Psi_n(y_1; (0, y_2, ..., y_n)).$$

Thus, from the very construction in the new variables the differential equations (A.1) will be expressed as:

$$\dot{y}_1 = 1,$$
$$\dot{y}_2 = 0,$$
$$\vdots \qquad\qquad\qquad\qquad\qquad\qquad\qquad\qquad\qquad\qquad\qquad\text{(A.19)}$$
$$\dot{y}_n = 0.$$

This is because, by the construction, y_1 equals to time t, so the derivative of it is equal to 1 and all the other variables y_k are equal to the values of the initial conditions of the solutions. Thus, they do not change along the solutions, so their time derivatives are equal to zero.

Finally, we can go back to the original variables $(x_1, x_2, ..., x_n)$ and get the change of variables:

$$\begin{aligned}
x_1 &= \Phi_1(y_1, y_2, \ldots, y_n), \\
x_2 &= \Phi_2(y_1, y_2, \ldots, y_n), \\
&\vdots \\
x_n &= \Phi_n(y_1, y_2, \ldots, y_n),
\end{aligned} \tag{A.20}$$

where the variables $(y_1, y_2, \ldots, y_n)$ satisfy the equations (A.19).

This finishes the sketch of the proof.

Some comments:

1) It may appear that the condition (A.11) was not used in the later part of the proof. Actually, it was used to make sure that each solution of the differential equations was going through just one point in the set of points in the form $(0, z_{0,2}, \ldots, z_{0,n})$. The derivative of the first variable, being non-zero, tells us that the zero will change to something else even under infinitesimal change of time.

2) Very often we will not have the explicit formula for (A.20). Notice that getting it explicitly would mean that we were able to explicitly solve the system (A.1). We know from practice that for many systems of equations of this kind the explicit solution is not known. Luckily, in many applications, we do not need to have explicit solutions. Just the fact that they exist may be sufficient.

APPENDIX B

THE DOUBLE FLOW BOX THEOREM

In this appendix, we will describe how, for given two systems of ODE (ordinary differential equations) that are commuting, there exists one set of variables, such that these variables are flow box variables for both ODE systems simultaneously.

1. Commuting ODE Systems

Assume that we have two systems of ordinary differential equations:

$$
\begin{aligned}
\frac{dx_1}{dt} &= f_1(x_1, x_2, \ldots, x_n), \\
\frac{dx_2}{dt} &= f_2(x_1, x_2, \ldots, x_n), \\
&\vdots \\
\frac{dx_n}{dt} &= f_n(x_1, x_2, \ldots, x_n),
\end{aligned}
\tag{B.1}
$$

and

$$
\begin{aligned}
\frac{dx_1}{d\tau} &= g_1(x_1, x_2, \ldots, x_n), \\
\frac{dx_2}{d\tau} &= g_2(x_1, x_2, \ldots, x_n), \\
&\vdots \\
\frac{dx_n}{d\tau} &= g_n(x_1, x_2, \ldots, x_n).
\end{aligned}
\tag{B.2}
$$

Then the systems (B.1) and (B.2) are called commuting if they satisfy:

$$\frac{d}{d\tau}f_1(x_1(\tau),x_2(\tau),...,x_n(\tau)) = \frac{d}{dt}g_1(x_1(t),x_2(t),...,x_n(t)),$$
$$\frac{d}{d\tau}f_2(x_1(\tau),x_2(\tau),...,x_n(\tau)) = \frac{d}{dt}g_2(x_1(t),x_2(t),...,x_n(t)),$$
$$\vdots \qquad\qquad\qquad\qquad\qquad\qquad\qquad\qquad\qquad\qquad (B.3)$$
$$\frac{d}{d\tau}f_n(x_1(\tau),x_2(\tau),...,x_n(\tau)) = \frac{d}{dt}g_n(x_1(t),x_2(t),...,x_n(t)),$$

where $(x_1(t), x_2(t),..., x_n(t))$ are solutions of the equations (B.1), and $(x_1(\tau), x_2(\tau),..., x_n(\tau))$ are solutions of the equations (B.2), as they pass through any point $(x_1, x_2,..., x_n)$. (This condition may possibly be satisfied only locally.)

2. The Double Flow Box Theorem for a Pair of Commuting ODE Systems

We will present a very brief sketch of the construction of the double flow box variables. The details of the construction are far outside of the scope of this text.

Similar to what we did in Appendix A, we assume that we start with a point $(x_{0,1}, x_{0,2},..., x_{0,n})$ such that not all functions $(f_1, f_2,..., f_n)$ are equal to zero at the point and such that not all functions $(g_1, g_2,..., g_n)$ are equal to zero there. We also assume that $(f_1, f_2,..., f_n)$ and $(g_1, g_2,..., g_n)$ at this point are not parallel in the vector sense, meaning that one cannot be obtained from the other by multiplication by a number.

Then, similarly to what we did in the Appendix A, by applying addition of constants and multiplication by a constant matrix to the original variables, we find the new variables $(z_1, z_2,..., z_n)$, such that our original starting point is now given by $(0, 0,..., 0)$.

Also, we expect that when the equations (B.1) and (B.2) are written in the new variables:

$$\frac{dz_1}{dt} = F_1(z_1, z_2, ..., z_n),$$

$$\frac{dz_2}{dt} = F_2(z_1, z_2, ..., z_n),$$

$$\vdots \tag{B.4}$$

$$\frac{dz_n}{dt} = F_n(z_1, z_2, ..., z_n),$$

and

$$\frac{dz_1}{d\tau} = G_1(z_1, z_2, ..., z_n),$$

$$\frac{dz_2}{d\tau} = G_2(z_1, z_2, ..., z_n),$$

$$\vdots \tag{B.5}$$

$$\frac{dz_n}{d\tau} = G_n(z_1, z_2, ..., z_n),$$

we have:

$$F_1(0, 0, ..., 0) = 1,$$
$$F_2(0, 0, ..., 0) = 0,$$
$$\vdots \tag{B.6}$$
$$F_n(0, 0, ..., 0) = 0,$$

and

$$G_1(0, 0, ..., 0) = 0,$$
$$G_2(0, 0, ..., 0) = 1,$$
$$G_3(0, 0, ..., 0) = 0,$$
$$\vdots \tag{B.7}$$
$$G_n(0, 0, ..., 0) = 0.$$

We will not prove this here, but the conditions (B.3) when expressed in the variables $(z_1, z_2, ..., z_n)$ become:

$$\frac{d}{d\tau} F_1(z_1(\tau), z_2(\tau), ..., z_n(\tau)) = \frac{d}{dt} G_1(z_1(t), z_2(t), ..., z_n(t)),$$
$$\frac{d}{d\tau} F_2(z_1(\tau), z_2(\tau), ..., z_n(\tau)) = \frac{d}{dt} G_2(z_1(t), z_2(t), ..., z_n(t)),$$
$$\vdots \qquad\qquad\qquad\qquad\qquad\qquad\qquad\qquad\qquad\qquad (B.8)$$
$$\frac{d}{d\tau} F_n(z_1(\tau), z_2(\tau), ..., z_n(\tau)) = \frac{d}{dt} G_n(z_1(t), z_2(t), ..., z_n(t)),$$

where $(z_1(t), z_2(t), ..., z_n(t))$ are the solutions of the equations (B.7), and $(z_1(\tau), z_2(\tau), ..., z_n(\tau))$ are solutions of the equations (B.8), as they pass through any point $(z_1, z_2, ..., z_n)$. (This condition may possibly be satisfied only locally.)

Now, we define the set of initial points expressed in the variables $(z_1, z_2, ..., z_n)$. This set which we will call A, is made of all points in a neighborhood of $(0, 0, ..., 0)$ that can be expressed as $(0, 0, z_3, z_4, ..., z_n)$.

Notice that because of (B.6) and (B.7), each solution of the equations (B.4) or (B.5) is crossing the set A at exactly one point.

Then, for any point $(z_1, z_2, ..., z_n)$ in a neighborhood of a point $(0, 0, ..., 0)$, we take a zig-zag path using the solutions of the equations (B.4) and (B.5), connecting the point $(z_1, z_2, ..., z_n)$ to a point from A. It can be shown that locally a given point $(z_1, z_2, ..., z_n)$ can be connected to only one point $(0, 0, z_{0,3}, z_{0,4}, ..., z_{0,n})$ in A. Also, we define t as the sum of all changes of the "time" parameters for the parts of the zig-zags along the solutions of (B.4), and we define τ as the sum of all changes of the "time" parameters for the parts of the zig-zags along the solutions of (B.5). Then we associate $(0, 0, z_{0,3}, z_{0,4}, ..., z_{0,n})$, t and τ, with the point $(z_1, z_2, ..., z_n)$, , getting:

$$z_1 = \Psi_1\left(t, \tau; (0, 0, z_{0,3}, ..., z_{0,n})\right),$$
$$z_2 = \Psi_2\left(t, \tau; (0, 0, z_{0,3}, ..., z_{0,n})\right),$$
$$\vdots \tag{B.9}$$
$$z_n = \Psi_n\left(t, \tau; (0, 0, z_{0,3}, ..., z_{0,n})\right).$$

Obviously, there is a question of getting different results for different zig-zag paths used. However, we may notice that, since we are using the solutions of the differential equations (B.3) and (B.5) for defining (B.9), we have, using (B.9) for arbitrary zig-zag's:

$$\frac{\partial z_i}{\partial t} = F_i,$$
$$\frac{\partial z_i}{\partial \tau} = G_i, \quad i = 1, ..., n. \tag{B.10}$$

Calculating the mixed derivatives, we then get:

$$\frac{\partial^2 z_i}{\partial \tau \partial t} = \frac{\partial F_i}{\partial \tau},$$
$$\frac{\partial^2 z_i}{\partial t \partial \tau} = \frac{\partial G_i}{\partial t}, \quad i = 1, ..., n. \tag{B.11}$$

Then, using the condition (B.8), we get:

$$\frac{\partial^2 z_i}{\partial \tau \partial t} = \frac{\partial^2 z_i}{\partial t \partial \tau}, \quad i = 1, ..., n. \tag{B.12}$$

This means that the mixed partial derivatives are equal, and this is the condition (compare to the Green's Theorem) for the independence from the path chosen, when defining a function z_i by using its partial derivatives. So, the formulas (B.9) do not depend on the path chosen to define z_i's

Now, we introduce new variables:

$$y_1 = t,$$
$$y_2 = \tau,$$
$$y_3 = z_{0,3}, \qquad (B.13)$$
$$\vdots$$
$$y_n = z_{0,n}.$$

From the construction we get:

$$\frac{dy_1}{dt} = 1,$$
$$\frac{dy_2}{dt} = 0,$$
$$\frac{dy_3}{dt} = 0, \qquad (B.14)$$
$$\vdots$$
$$\frac{dy_n}{dt} = 0.$$

and

$$\frac{dy_1}{d\tau} = 0,$$
$$\frac{dy_2}{d\tau} = 1,$$
$$\frac{dy_3}{d\tau} = 0, \qquad (B.15)$$
$$\vdots$$
$$\frac{dy_n}{d\tau} = 0.$$

Finally, we can go back to the original variables $(x_1, x_2, ..., x_n)$, getting the change of variables:

$$\begin{aligned} x_1 &= \Phi_1(y_1, y_2, ..., y_n), \\ x_2 &= \Phi_2(y_1, y_2, ..., y_n), \\ &\vdots \\ x_n &= \Phi_n(y_1, y_2, ..., y_n), \end{aligned} \quad (B.16)$$

where the variables $(y_1, y_2, ..., y_n)$ satisfy the equations (B.14) and (B.15).

This finishes the sketch of the construction.

APPENDIX C

THE DARBOUX THEOREM, THE VERSION FOR THE POISSON BRACKETS

The Darboux theorem is easy to find in literature, but it is usually given and proven in the language of differential forms. Since this book is using the language of the Poisson Brackets, we want to present the version of the Darboux Theorem and a sketch of its proof that uses the Poisson Brackets.

Earlier, in Chapter VII, we defined Poisson Brackets using the positions and momenta obtained from a given constrained Lagrangian. However, Poisson Brackets can be defined in a more general way.

Assume we have a space described locally by $2N$ variables $(x_1, x_2, ..., x_{2N})$. For an ordered pair of functions $F = F(x_1, x_2, ..., x_{2N})$ and $G = G(x_1, x_2, ..., x_{2N})$, we define a new function of $(x_1, x_2, ..., x_{2N})$, called Poisson Brackets and denoted by $\{F, G\}$. The function $\{F, G\}$ is given as:

$$\{F, G\} = \sum_{i=1}^{2N} \sum_{j=1}^{2N} \frac{\partial F}{\partial x_i} \cdot P_{ij} \cdot \frac{\partial G}{\partial x_j} =$$

$$= \left(\frac{\partial F}{\partial x_1}, \frac{\partial F}{\partial x_2}, ..., \frac{\partial F}{\partial x_{2N}} \right) \begin{pmatrix} P_{11} & P_{12} & \cdots & P_{1,2N} \\ P_{21} & P_{22} & \cdots & P_{2,2N} \\ \vdots & \vdots & & \vdots \\ P_{2N,1} & P_{2N,2} & \cdots & P_{2N,2N} \end{pmatrix} \begin{pmatrix} \frac{\partial G}{\partial x_1} \\ \frac{\partial G}{\partial x_2} \\ \vdots \\ \frac{\partial G}{\partial x_{2N}} \end{pmatrix}, \quad (C.1)$$

where each entry in the matrix P_{ij} in (C.1) is a smooth enough function of $(x_1, x_2, ..., x_{2N})$. We assume that for each $(x_1, x_2, ..., x_{2N})$, the matrix P_{ij} is invertible and antisymmetric, meaning that we have $P_{ij}(x_1, x_2, ..., x_{2N}) = -P_{ji}(x_1, x_2, ..., x_{2N})$. We also assume that the matrix P_{ij} is such that the Poisson Brackets (C.1) satisfy the Jacobi Identity:

$$\{\{F,G\},H\}+\{\{G,H\},F\}+\{\{H,F\},G\}=0 \text{ for any } (x_1,x_2,...,x_{2N}). \tag{C.2}$$

Notice that by using $F = x_i$ and $G = x_j$ in (C.1), we get $P_{ij} = \{x_i, x_j\}$. So, (C.1) can be rewritten as:

$$\{F,G\} = \sum_{i=1}^{2N}\sum_{j=1}^{2N}\frac{\partial F}{\partial x_i}\cdot\{x_i,x_j\}\cdot\frac{\partial G}{\partial x_j} =$$

$$= \left(\frac{\partial F}{\partial x_1},\frac{\partial F}{\partial x_2},...,\frac{\partial F}{\partial x_{2N}}\right)\begin{pmatrix}\{x_1,x_1\} & \{x_1,x_2\} & \cdots & \{x_1,x_{2N}\}\\ \{x_2,x_1\} & \{x_2,x_2\} & \cdots & \{x_2,x_{2N}\}\\ \vdots & \vdots & & \vdots\\ \{x_{2N},x_1\} & \{x_{2N},x_2\} & \cdots & \{x_{2N},x_{2N}\}\end{pmatrix}\begin{pmatrix}\frac{\partial G}{\partial x_1}\\ \frac{\partial G}{\partial x_2}\\ \vdots\\ \frac{\partial G}{\partial x_{2N}}\end{pmatrix}. \tag{C.3}$$

Notice also that if we choose any function $H = H(x_1, x_2, ..., x_{2N})$ and define a set of ordinary differential equations for the basic variables as:

$$\begin{aligned}\dot{x}_1 &= \{x_1,H\},\\ \dot{x}_2 &= \{x_2,H\},\\ &\vdots\\ \dot{x}_{2N} &= \{x_{2N},H\},\end{aligned} \tag{C.4}$$

then we also have:

$$\dot{F} = \sum_{i=1}^{2N}\frac{\partial F}{\partial x_i}\cdot\dot{x}_i = \sum_{i=1}^{2N}\frac{\partial F}{\partial x_i}\cdot\{x_i,H\} = \sum_{i=1}^{2N}\frac{\partial F}{\partial x_i}\cdot\left[\sum_{j=1}^{2N}\sum_{k=1}^{2N}\frac{\partial x_i}{\partial x_j}\cdot\{x_j,x_k\}\cdot\frac{\partial H}{\partial x_k}\right] =$$

$$= \sum_{i=1}^{2N}\frac{\partial F}{\partial x_i}\cdot\left[\sum_{j=1}^{2N}\sum_{k=1}^{2N}\delta_{jk}\cdot\{x_j,x_k\}\cdot\frac{\partial H}{\partial x_k}\right] = \sum_{i=1}^{2N}\frac{\partial F}{\partial x_i}\cdot\left[\sum_{j=1}^{2N}\sum_{k=1}^{2N}\delta_{ij}\cdot\{x_j,x_k\}\cdot\frac{\partial H}{\partial x_k}\right] =$$

$$= \sum_{i=1}^{2N}\frac{\partial F}{\partial x_i}\cdot\left[\sum_{k=1}^{2N}\{x_i,x_k\}\cdot\frac{\partial H}{\partial x_k}\right] = \sum_{i=1}^{2N}\sum_{k=1}^{2N}\frac{\partial F}{\partial x_i}\cdot\{x_i,x_k\}\cdot\frac{\partial H}{\partial x_k} = \{F,H\}.$$

In the above we used the Kronecker delta δ_{ij}, defined as:

$$\delta_{ij} = \begin{cases} 1 & \text{when } i = j \\ 0 & \text{when } i \neq j \end{cases} \quad i, j = 1, ..., N. \tag{C.5}$$

So, from the general properties of the Poisson Brackets, if we have the equations (C.4), we automatically get:

$$\dot{F} = \{F, H\}. \tag{C.6}$$

Also, since the Poisson Brackets are defined using derivatives, we get:

$$\{FG, H\} = F\{G, H\} + G\{F, H\}, \tag{C.7}$$

which is a simple result of applying the power rule for derivatives in (C.1).

The Darboux Theorem can now be formulated as follows:

Assume we have the Poisson Brackets defined as above. Then, in a neighborhood of any given point, there exists a change of variables such that new variables, denoted by $(Q_1, Q_2, ..., Q_N, P_1, P_2, ..., P_N)$, are satisfying:

$$\begin{aligned} \{Q_i, Q_j\} &= 0, \\ \{P_i, P_j\} &= 0, \\ \{Q_i, P_j\} &= \delta_{ij}, \quad i, j = 1, ..., N. \end{aligned} \tag{C.8}$$

In other words, for any Poisson Brackets, there always exist variables $(Q_1, Q_2, ..., Q_N, P_1, P_2, ..., P_N)$ that are canonical.

Sketch of the proof:

Assume we have Poisson Brackets $\{F,G\}$ on variables $(x_1, x_2, ..., x_{2N})$ satisfying the conditions above. We will describe a construction of variables $(Q_1, Q_2, ..., Q_N, P_1, P_2, ..., P_N)$ that satisfy the conditions (C.8).

The construction starts with defining P_1 as:

$$P_1 = x_1. \tag{C.9}$$

Then, we define the set of differential equations:

$$\begin{aligned}
\dot{x}_1 &= \{x_1, P_1\} = \{x_1, x_1\} = 0, \\
\dot{x}_2 &= \{x_2, P_1\} = \{x_2, x_1\}, \\
\dot{x}_3 &= \{x_3, P_1\} = \{x_3, x_1\}, \\
&\vdots \\
\dot{x}_{2N} &= \{x_{2N}, P_1\} = \{x_{2N}, x_1\},
\end{aligned} \tag{C.10}$$

where a dot above variables means derivatives with respect to an outside parameter. We can write (C.10) as:

$$\begin{aligned}
\dot{x}_1 &= 0, \\
\dot{x}_2 &= f_2(x_1, x_2, ..., x_{2N}), \\
\dot{x}_3 &= f_3(x_1, x_2, ..., x_{2N}), \\
&\vdots \\
\dot{x}_{2N} &= f_{2N}(x_1, x_2, ..., x_{2N}),
\end{aligned} \tag{C.11}$$

where the functions on the right side are just a way to write the Poisson Brackets from (C.10). In a way that makes the variables $(x_1, x_2, ..., x_{2N})$ showing up explicitly.

Since $\dot{x}_1 = 0$, then $x_1 = const$ for any solution of (C.11). So, if we choose a particular x_1, for that x_1 we can define:

$$g_2(x_2, x_3, ..., x_{2N}) = f_2(x_1, x_2, x_3, ..., x_{2N}),$$
$$g_3(x_2, x_3, ..., x_{2N}) = f_3(x_1, x_2, x_3, ..., x_{2N}),$$
$$\vdots \qquad (C.12)$$
$$g_{2N}(x_2, x_3, ..., x_{2N}) = f_{2N}(x_1, x_2, x_3, ..., x_{2N}).$$

Placing (C.12) into (C.11) and ignoring the first entry in (C.11), we get:

$$\dot{x}_2 = g_2(x_2, x_3, ..., x_{2N}),$$
$$\dot{x}_3 = g_3(x_2, x_3, ..., x_{2N}),$$
$$\vdots \qquad (C.13)$$
$$\dot{x}_{2N} = g_{2N}(x_2, x_3, ..., x_{2N}).$$

Not all entries on the right sides of the equations (C.10) are equal to zero, because the right sides of (C.10) make a column in the Poisson Brackets matrix. This matrix is invertible, so it cannot have entire column made of all zeros. Therefore, not all entries on the right side of (C.13) are zeros. So, we can use the flow box theorem (Appendix A), which tells us that locally there exists such a coordinate system (it is convenient to index the new variables starting at 2):

$$z_2 = z_2(x_2, x_3, ..., x_n),$$
$$z_3 = z_3(x_2, x_3, ..., x_n),$$
$$\vdots \qquad (C.14)$$
$$z_{2N} = z_{2N}(x_2, x_3, ..., x_n),$$

such that
$$\dot{z}_2 = 1,$$
$$\dot{z}_3 = 0,$$
$$\vdots \qquad (C.15)$$
$$\dot{z}_{2N} = 0.$$

Using (C.6) in (C.15), we get:

$$\dot{z}_2 = \{z_2, P_1\} = 1,$$
$$\dot{z}_3 = \{z_3, P_1\} = 0,$$
$$\vdots \qquad (C.16)$$
$$\dot{z}_{2N} = \{z_{2N}, P_1\} = 0.$$

This construction replaces (locally) the original coordinate system $(x_1, x_2, ..., x_{2N})$ with the coordinate system $(z_2, P_1, z_3, ..., z_{2N})$.

For convenience, let us temporary introduce:

$$z_1 = P_1. \qquad (C.17)$$

Since from (C.16) and (C.17) we have $1 = \{z_2, z_1\}$, we also have, for any z_i, $i = 1, ..., 2N$:

$$\{\{z_2, z_1\}, z_i\} = 0. \qquad (C.18)$$

The (C.18) is true because the Poisson Brackets, in any variables, are defined by derivatives, as in (C.3), and all derivatives of 1 are equal to zero.

The Jacobi Identity gives:

$$\{\{z_2, z_1\}, z_i\} + \{\{z_1, z_i\}, z_2\} + \{\{z_i, z_2\}, z_1\} = 0$$
$$(C.19)$$

Using (C.18) in (C.19), we get:

$$\{\{z_1, z_i\}, z_2\} + \{\{z_i, z_2\}, z_1\} = 0.$$

$$\{\{z_i, z_2\}, z_1\} = -\{\{z_1, z_i\}, z_2\}.$$

$$\{\{z_i, z_2\}, z_1\} = \{\{z_i, z_1\}, z_2\}.$$

$$\{\{z_2, z_i\}, z_1\} = \{z_2, \{z_i, z_1\}\}. \tag{C.20}$$

Now, let us define two systems of ordinary differential equations.

The first system is defined as:

$$\begin{aligned}
\frac{dz_1}{dt} &= \{z_1, z_1\} = 0, \\
\frac{dz_2}{dt} &= \{z_2, z_1\} = 1, \\
\frac{dz_3}{dt} &= \{z_3, z_1\}, \\
&\vdots \\
\frac{dz_{2N}}{dt} &= \{z_{2N}, z_1\}.
\end{aligned} \tag{C.21}$$

The second system is defined as:

$$\begin{aligned}
\frac{dz_1}{d\tau} &= \{z_2, z_1\} = 1, \\
\frac{dz_2}{d\tau} &= \{z_2, z_2\} = 0, \\
\frac{dz_3}{d\tau} &= \{z_2, z_3\}, \\
&\vdots \\
\frac{dz_{2N}}{d\tau} &= \{z_2, z_{2N}\}.
\end{aligned} \tag{C.22}$$

Rewrite the equations (C.21) and (C.22) as:

$$\frac{dz_1}{dt} = F_1(z_1, z_2, ..., z_n),$$
$$\frac{dz_2}{dt} = F_2(z_1, z_2, ..., z_n),$$
$$\vdots$$
$$\frac{dz_n}{dt} = F_n(z_1, z_2, ..., z_n),$$
(C.23)

and

$$\frac{dz_1}{d\tau} = G_1(z_1, z_2, ..., z_n),$$
$$\frac{dz_2}{d\tau} = G_2(z_1, z_2, ..., z_n),$$
$$\vdots$$
$$\frac{dz_n}{d\tau} = G_n(z_1, z_2, ..., z_n),$$
(C.24)

where

$$F_1(z_1, z_2, ..., z_n) = \{z_1, z_1\},$$
$$F_2(z_1, z_2, ..., z_n) = \{z_2, z_1\},$$
$$\vdots$$
$$F_n(z_1, z_2, ..., z_n) = \{z_{2N}, z_1\},$$
(C.25)

and

$$G_1(z_1, z_2, ..., z_n) = \{z_2, z_1\},$$
$$G_2(z_1, z_2, ..., z_n) = \{z_2, z_2\},$$
$$\vdots$$
$$G_n(z_1, z_2, ..., z_n) = \{z_2, z_{2N}\}.$$
(C.26)

Placing (C.25) and (C.26) into (C.20), we get:

$$\{G_i, z_1\} = \{z_2, F_i\}.$$
(C.27)

Then, using (C.6), we get:

$$\frac{dG_i}{dt} = \frac{dF_i}{d\tau}. \tag{C.28}$$

Condition (C.28) is the same as condition (B.8) from Appendix B of commuting systems of equations. So, the equations (C.21) and (C.22) commute.

Then, the result (B.9) of Appendix B tell us that there exists a change of variables that can be written as:

$$\begin{aligned}
z_1 &= \Psi_1\left(t, \tau; \left(0, 0, z_{0,3}, ..., z_{0,n}\right)\right), \\
z_2 &= \Psi_2\left(t, \tau; \left(0, 0, z_{0,3}, ..., z_{0,n}\right)\right), \\
&\vdots \\
z_n &= \Psi_n\left(t, \tau; \left(0, 0, z_{0,3}, ..., z_{0,n}\right)\right).
\end{aligned} \tag{C.28}$$

Then, since the derivative of z_1 with respect to t is equal to 1, then z_1 is equal to t up to an additive constant, so t can be expressed by z_1 in (C.28). Similarly, τ can be replaced by z_2 in (C.28). So, we get:

$$\begin{aligned}
z_1 &= \Psi_1\left(z_1, z_2; \left(0, 0, z_{0,3}, ..., z_{0,n}\right)\right), \\
z_2 &= \Psi_2\left(z_1, z_2; \left(0, 0, z_{0,3}, ..., z_{0,n}\right)\right), \\
&\vdots \\
z_n &= \Psi_n\left(z_1, z_2; \left(0, 0, z_{0,3}, ..., z_{0,n}\right)\right).
\end{aligned} \tag{C.29}$$

Introducing new variables $(y_3, ..., y_n)$, we get:

$$z_1 = z_1,$$
$$z_2 = z_2,$$
$$y_3 = z_{0,3}, \tag{C.30}$$
$$\vdots$$
$$y_n = z_{0,n}.$$

We obtain new variables $(z_1, z_2, y_3, ..., y_n)$ that replace the original variables $(x_2, x_3, ..., x_{2N})$. In these new variables, as shown in Appendix B, we have:

$$\frac{dz_1}{dt} = 0,$$
$$\frac{dz_2}{dt} = 1,$$
$$\frac{dy_3}{dt} = 0,$$
$$\frac{dy_4}{dt} = 0, \tag{C.31}$$
$$\vdots$$
$$\frac{dy_n}{dt} = 0,$$

and

$$\frac{dz_1}{d\tau} = 1,$$
$$\frac{dz_2}{d\tau} = 0,$$
$$\frac{dy_3}{d\tau} = 0,$$
$$\frac{dy_4}{d\tau} = 0, \tag{C.32}$$
$$\vdots$$
$$\frac{dy_n}{d\tau} = 0.$$

Knowing that the equations (C.31) and (C.32) can be written using the Poisson Brackets on the right sides and comparing with earlier equations, we get:

$$\{z_2, z_1\} = 1,$$
$$\{z_1, y_i\} = 0, \qquad (C.33)$$
$$\{z_2, y_i\} = 0, \quad i = 3, ..., 2N.$$

Going back to the variables $P_1 = z_1$ as in formula (C.17) and defining $Q_1 = z_2$ we get:

$$\{Q_1, P_1\} = 1,$$
$$\{Q_1, y_i\} = 0, \qquad (C.34)$$
$$\{P_1, y_i\} = 0, \quad i = 3, ..., 2N.$$

So, we have two variables that satisfy $\{Q_1, P_1\} = 1$, and they have Poisson Brackets equal to zero with all other variables. Then, we can start the process anew with the variables $(y_3, ..., y_n)$ getting Q_2 and P_2 and so on, until we have $(Q_1, Q_2, ..., Q_N, P_1, P_2, ..., P_N)$ that satisfy conditions (3.8).

Getting Q_2 and P_2 that give a zero Poisson Brackets with Q_1 and P_1 is the result of the property of the Poisson Brackets that says that if F has a zero Poisson Brackets with variables $(y_3, ..., y_n)$, then F has the Poisson Brackets with any function of $(y_3, ..., y_n)$.

This ends the sketch of the proof of the Darboux theorem.

Important note: Please notice that at the end of the construction, the P_1 still equals to x_1, defined in (C.9) at the very beginning of the construction. It means that P_1 can be chosen arbitrarily, as long as it is a variable in an arbitrarily chosen variable system.

APPENDIX D

THE DARBOUX THEOREM, THE VERSION WITH CONSTRAINTS, AS APPLIED TO DIRAC'S BRACKETS

Now, we will show the version of the Darboux theorem that we use in Chapter XIV.

Assume we have variables $(x_1, x_2, ..., x_{2N})$. (These variables may include positions and momenta.) Now, also assume that some variables are the constraints of Class II that we want to use to define the Dirac's Brackets. Without losing generality, assume that the constraints are at the beginning of the list of the variables above. So, we have variables written as $(\chi_1, \chi_2, ..., \chi_{2K}, x_{2K+1}, x_{2K+2}, ..., x_{2N})$.

We start by defining the variable $\phi_{P_1} = \chi_1$. Then, using what we did in Appendix A, but only for variables $(\chi_1, \chi_2, ..., \chi_{2K})$, we get such a variable ϕ_{Q_1} such that $\{\phi_{Q_1}, \phi_{P_1}\} = 1$. Since we only used the constraints $(\chi_1, \chi_2, ..., \chi_{2K})$ to construct ϕ_{Q_1}, then ϕ_{Q_1} is a constraint, up to an additive constant. Then, we can include this additive constant in the definition of ϕ_{Q_1}, without changing its Poisson Brackets.

Then, using the Appendix B for ϕ_{Q_1} and ϕ_{P_1}, we obtain new variables $(\phi_{Q_1}, \phi_{P_1}, z_3, z_4, ..., z_{2N})$. They are such that:

$$\begin{aligned} \{\phi_{Q_1}, \phi_{P_1}\} &= 1, \\ \{\phi_{Q_1}, z_i\} &= 0, \\ \{\phi_{P_1}, z_i\} &= 0, \quad i = 3, ..., 2N. \end{aligned} \qquad (D.1)$$

Then, we take another change of variables by using the remaining constraints as new variables. We get the variables denoted as $(\phi_{Q_1}, \phi_{P_1}, \phi_3, \phi_4, ..., \phi_{2K}, y_{2K+1}, y_{2K+2}, ..., y_{2N})$, satisfying:

$$\{\phi_{Q_1}, \phi_{P_1}\} = 1,$$
$$\{\phi_{Q_1}, \phi_j\} = 0, \quad j = 3, \ldots, 2K,$$
$$\{\phi_{P_1}, \phi_j\} = 0, \quad j = 3, \ldots, 2K, \quad \text{(D.2)}$$
$$\{\phi_{Q_1}, y_i\} = 0, \quad i = 2K+1, \ldots, 2N,$$
$$\{\phi_{P_1}, y_i\} = 0, \quad i = 2K+1, \ldots, 2N,$$

Then, we repeat the process getting two more constraints ϕ_{Q_2} and ϕ_{P_2} with analogous properties. We will continue until all constraints are covered. Then we continue with the rest of variables, but now including all of them as in the steps used in Appendix A.

After the process is done, we have the variables:

$$\left(\phi_{Q_1}, \phi_{P_1}, \phi_{Q_2}, \phi_{P_2}, \ldots, \phi_{Q_K}, \phi_{P_K}, Q_{K+1}, P_{K+1}, \ldots, Q_N, P_N\right),$$

such that:

$$\{\phi_{Q_i}, \phi_{P_j}\} = \delta_{ij}, \quad i, j = 1, \ldots, K,$$
$$\{Q_i, P_j\} = \delta_{ij}, \quad i, j = K+1, \ldots, N,$$
$$\{\phi_{Q_i}, Q_j\} = 0, \quad i = 1, \ldots, K, \quad j = K+1, \ldots, N,$$
$$\{\phi_{P_i}, Q_j\} = 0, \quad i = 1, \ldots, K, \quad j = K+1, \ldots, N, \quad \text{(D.3)}$$
$$\{\phi_{Q_i}, P_j\} = 0, \quad i = 1, \ldots, K, \quad j = K+1, \ldots, N,$$
$$\{\phi_{P_i}, P_j\} = 0, \quad i = 1, \ldots, K, \quad j = K+1, \ldots, N.$$

So, the variables $\left(\phi_{Q_1}, \phi_{P_1}, \phi_{Q_2}, \phi_{P_2}, \ldots, \phi_{Q_K}, \phi_{P_K}, Q_{K+1}, P_{K+1}, \ldots, Q_N, P_N\right)$ are canonical variables, such that the first $2K$ of them are the constraints of Class II.

We may then decide to use these constraints to define the Dirac's Brackets. We use the variables in this way in Chapter XIV.

APPENDIX E

THE DARBOUX-BOX VARIABLES

Now we will show the theorem that tells us that it is possible to combine the box variables with a version of the Darboux Theorem. This theorem is sometimes referred to as the Carathéodory–Jacobi–Lie theorem. We would rather use the names Darboux-Box theorem and Darboux-Box variables.

As before, assume we have a system described by general variables $(x_1, x_2, ..., x_{2N})$. These variables are not just positions of the system, but they represent all variables describing the system. So they possibly include positions, velocities, momenta, and any possible combinations of them.

Assume also that the dynamics of the system is given by a Hamiltonian $H = H(x_1, x_2, ..., x_{2N})$ and some Poisson Brackets. By the Poisson Brackets, we mean the brackets defined in Appendix C, where the Poisson Brackets were defined independently of the existence of a Lagrangian.

Assume that we have a point $(x_{0,1}, x_{0,2}, ..., x_{0,2N})$ such that at least one time derivative of $(x_1, x_2, ..., x_{2N})$ is not zero at the point, if the time derivatives are produced by the given Hamiltonian and the given Poisson Brackets.

The construction below is local, it is defined on an open neighborhood of the point.

Since the time derivatives of variables $(x_1, x_2, ..., x_{2N})$ are given by the generalized Poisson Brackets of variables with the Hamiltonian H, and the generalized Poisson Brackets are defined using derivatives, at least one partial derivative of the Hamiltonian H, say the one with respect to x_1, must be different than zero at $(x_{0,1}, x_{0,2}, ..., x_{0,2N})$. Then, the Hamiltonian H can replace x_1, and the variables $(H, x_2, ..., x_{2N})$ still make a good local variable system.

Then, we can use the procedure described in Appendix C to get the canonical variables. If we use H as the starting variable P_1 in this procedure, we will still have the variable H at the end of this procedure. According to the Appendix C, in the neighborhood of the point $(x_{0,1}, x_{0,2}, ..., x_{0,2N})$, we end up with the canonical variables $(Q_1, P_1, Q_2, P_2, ..., Q_N, P_N)$, such that:

$$\{Q_i, Q_j\} = 0,$$
$$\{P_i, P_j\} = 0, \qquad\qquad\qquad\qquad\qquad\qquad (E.1)$$
$$\{Q_i, P_j\} = \delta_{ij}, \quad i, j = 1, ..., N.$$

We also have:

$$P_1 = H.$$

In other words, for any Hamiltonian dynamical system, we can locally find coordinates that are canonical coordinates and such that the Hamiltonian is the first momentum variable.

Such variables we will call the Darboux-box variables.

We do not use this result in Volume I of this text. We use it in Volumes II and III.

APPENDIX F

CALCULATIONAL SHORTCUTS FOR DIRAC'S BRACKETS

In this appendix, we present shortcuts that, in some situations, may significantly shorten the calculations associated with Dirac's Brackets.

1. Dirac Brackets of a constraint used to define these Dirac Brackets with any function F are equal to 0 on constraints.

$$\{F, \chi_k\}_D = 0 \tag{F.1}$$

This is because the vector on the very right side of the definition of the Dirac's Brackets is one of the vectors used to define the constraint matrix. Then, it is multiplied by the inverse of that matrix, so we get a vertical vector with all zeros and just one 1. The 1 is in such a place that we exactly get the part that subtracts from the expression on the left side of the definition of Dirac's Brackets. Therefore, we get:

$$\{F, \chi_k\}_D = \{F, \chi_k\} - \{F, \chi_k\} = 0 \tag{F.2}$$

2. Any function G of constraints used to define the Dirac Brackets has zero Dirac Brackets with any variable F.

Notice that a function G of some constraints used to define the Dirac's Brackets is equal to a constant on all points that satisfy all constraints. Then by addition of a constant, which does not change the derivatives and therefore does not change the values of the Dirac's Brackets, we can make this function to be equal to zero on all points satisfying the constraints that define the Dirac's Brackets. Therefore, following the results of Chapter VI, such a function can be written as a linear combination of the constraints, plus a constant (which can be ignored when calculating the Dirac's Brackets):

$$G = c + \sum_{k=1}^{M} u_k \chi_k \tag{F.3}$$

Then, following the general properties of the Dirac's Brackets, we get:

$$\{G,F\} = \left\{c + \sum_{k=1}^{M} u_k \chi_k, F\right\} = \left\{\sum_{k=1}^{M} u_k \chi_k, F\right\} = \sum_{k=1}^{M} \{u_k \chi_k, F\} =$$
$$= \sum_{k=1}^{M} \left[u_k \{\chi_k, F\} + \chi_k \{u_k, F\}\right] = \sum_{k=1}^{M} \left[u_k \cdot 0 + 0 \cdot \{u_k, F\}\right] = 0.$$
(F.4)

3. If a quantity F has a zero Poisson Brackets with all constraints used to define the Dirac's Brackets, then its Dirac Bracket with all other quantities is equal to its Poisson Brackets with the same quantity:

$$\{F,G\}_D = \{F,G\}$$
(F.5)

This is because the horizontal vector in the definition of the Dirac Brackets will be made of all zeros. So, the second part of the definition will be equal to zero.

4. We can replace quantities using constraints used to define the Dirac's Brackets, and the Dirac's Brackets will not change as a result.

This is because in Chapter VI, we have shown that such substitutions are equivalent to adding linear combinations of constraints and, as shown in point 3 above, have zero Dirac's Brackets with all variables.

5. Results for Dirac's Brackets for special form of constraints.

Starting with canonical variables, $(q_1, ..., q_N, p_1, ..., p_N)$, the constraints in the form:

$$\begin{aligned} \theta_{1r} &= q_r - q_r(q_1, ..., q_R, p_1, ..., p_R) = 0, \\ \theta_{2r} &= p_r = 0, \end{aligned} \qquad r = R+1, ..., N.$$
(F.6)

give as their Dirac's Brackets:

$$\{q_i, q_j\}_D = 0,$$

(continued in the next page)

$$\{p_i, p_j\}_D = 0,$$
$$\{q_i, p_j\}_D = \delta_{ij}, \quad i, j = 1, \ldots, R. \tag{F.7}$$

Consequently, the Dirac's Brackets of the quantities $F = F(q_1, \ldots, q_R, p_1, \ldots, p_R)$ and $G = G(q_1, \ldots, q_R, p_1, \ldots, p_R)$ will be given as:

$$\{F, G\}_D = \sum_{i=1}^{R} \left[\frac{\partial F}{\partial q_i} \cdot \frac{\partial G}{\partial p_i} - \frac{\partial F}{\partial p_i} \cdot \frac{\partial G}{\partial q_i} \right], \tag{F.8}$$

which has a form identical to the regular Poisson Brackets but is using only the variables $(q_1, \ldots, q_R, p_1, \ldots, p_R)$.

Then, we can use the constraints (F.6) to substitute $q_r = q_r(q_1, \ldots, q_R, p_1, \ldots, p_R)$ and $p_r = 0$, $r = R+1, \ldots, N$, into all functions that contain q_r, p_r, $r = R+1, \ldots, N$.

We know that such substitutions do not change the results of calculations involving the Dirac's Brackets. Combining this with the fact that Dirac's Brackets can be used instead of the Poisson Brackets to obtain the Hamilton's equations of motion, we conclude that we can eliminate all variables q_r, p_r, $r = R+1, \ldots, N$, and still reproduce the original equations of motion.

So, we conclude that if we have the constraints in the form (F.6), we do not need to perform any complicated calculations to obtain the Dirac's Brackets, since they are given in a simple form (F.8). Then, the properties of the Dirac's Brackets are such that even if we need to calculate them for the functions that still contain q_r, p_r, $r = R+1, \ldots, N$, we can use the form (F.8). We just need to eliminate q_r, p_r, by substituting $q_r = q_r(q_1, \ldots, q_R, p_1, \ldots, p_R)$ and $p_r = 0$ into these functions, before we use (F.8).

In the sense described above, the variables q_r, p_r, $r = R+1, \ldots, N$, are completely eliminated from the formalism by the constraints (F.6) and the use of the Dirac's Brackets.

APPENDIX G

AN INDEPENDENT DEFINITION OF THE POISSON BRACKETS

The Poisson Brackets are usually defined in relation to the existing Lagrangian of the system. However, quite often we need to define the Poisson Brackets independently of the Lagrangian. This is especially true when starting with the equations of motion rather than a Lagrangian when defining a mechanical system. So, here we present how the Poisson Brackets may be defined without a Lagrangian. Often, they are called Generalized Poisson Brackets. We decided to still call them the Poisson Brackets.

Assume we have a system described by general variables $(w_1, w_2, ..., w_{2N})$. These variables are not just positions of the system; they represent all variables describing the system. So they possibly include positions, velocities, momenta, and any possible combinations of them. For shortness, in the text below, we will often denote $(w_1, w_2, ..., w_{2N})$ by (w).

To define Generalized Poisson Brackets, we may start with defining a matrix field by specifying, at every point (w), a $2N \times 2N$ matrix $P_{mk}(w_1, w_2, ..., w_{2N})$, $m, k = 1, ..., 2N$.

We require that the matrix satisfies the following conditions at every point (w):

$$\det P_{mk}(w) \neq 0, \qquad \text{(invertibility)} \qquad \text{(G.1)}$$

$$P_{km}(w) = -P_{mk}(w), \qquad \text{(antisymmetry)} \qquad \text{(G.2)}$$

$$\sum_{m=1}^{2N} \left[P_{ma}(w) \cdot \frac{\partial P_{bc}(w)}{\partial w_m} + P_{mb}(w) \cdot \frac{\partial P_{ca}(w)}{\partial w_m} + P_{mc}(w) \cdot \frac{\partial P_{ab}(w)}{\partial w_m} \right] = 0. \quad \text{(Jacobi condition)} \quad \text{(G.3)}$$

Having $P_{mk}(w_1, w_2, ..., w_{2N})$ given as above, we define an operator denoted by $\{\ ,\ \}$, called the Poisson Brackets, that for two functions $F(w)$ and $G(w)$ assigns another function of (w), given as:

$$\{F,G\} = \sum_{m,k=1}^{2N} \left[\frac{\partial F(w)}{\partial w_m} \cdot P_{mk}(w) \cdot \frac{\partial G(w)}{\partial w_k} \right]. \tag{G.4}$$

We should stress that $\{F,G\}$ is a function of (w). We assume that when we change the variables, the function $\{F,G\}$ changes by just expressing old variables (w) in $\{F,G\}$ by the new variables.

It is also important to realize that $P_{mk}(w)$ in (G.4) are not regular functions, in the sense that when we change the variables, the expressions for P_{mk} in the new variables are not obtained by simply expressing old variables (w) in each P_{mk} by the new variables. Actually, if we want $\{F,G\}$ to transform as the usual functions, and we do, we must require that the transformation of P_{mk} is given by:

$$P_{mk}(z) = \sum_{i,j=1}^{2N} \left[\frac{\partial z_m}{\partial w_i} \cdot P_{ij}(w) \cdot \frac{\partial z_k}{\partial w_j} \right]_{w=w(z)}, \quad m,k=1,...,2N. \tag{G.5}$$

The details leading to (G.5) can be found in Chapter XXV, in Volume III, in the calculations leading to (25.28).

The Poisson Brackets defined by (G.4) have the following properties:

1) $\{F,G\} = -\{G,F\}$, (antisymmetry) (G.6)

2) $\{c_1 F_1 + c_2 F_2, G\} = \{c_1 F_1, G\} + \{c_2 F_2, G\}$, (linearity) (G.7)

where c_1 and c_2 are constants,

3) $\{F_1 \cdot F_2, G\} = F_1 \cdot \{F_2, G\} + \{F_1, G\} \cdot F_2$, (product rule) (G.8)

4) $\{F, G_1 + G_2\} = G_1 \cdot \{F, G_2\} + \{F, G_1\} \cdot G_2$, (product rule) (G.9)

5) $\{\{F,G\},H\} + \{\{G,H\},F\} + \{\{H,F\},G\} = 0,$ (Jacobi Identity) (G.10)

where $F, G,$ and H are functions of (w),

6) If $\{F,G\} = 0$ for all functions $G(w)$, then $F(w)$ must be a constant function.

7) $P_{mk}(w) = \{w_m, w_k\}, \quad m, k = 1, \ldots, 2N,$ (G.11)

8) $\{F,G\} = \sum_{m,k=1}^{2N} \left[\dfrac{\partial F(w)}{\partial w_m} \cdot \{w_m, w_k\} \cdot \dfrac{\partial G(w)}{\partial w_k} \right],$ (G.12)

9) The following two equalities hold:

$\{F,G\} = \sum_{m=1}^{2N} \left[\{F, w_m\} \cdot \dfrac{\partial G}{\partial w_m} \right],$ (G.13)

and

$\{F,G\} = \sum_{m=1}^{2N} \left[\dfrac{\partial F}{\partial w_m} \cdot \{w_m, G\} \right].$ (G.14)

The properties 1) to 9), are quite easy to obtain from the definition (G.1) to (G.4).

APPENDIX H

THE LEGENDRE TRANSFORMATIONS IN NON-TYPICAL VARIABLES

In Chapter III, Volume I, we showed how to obtain the Hamiltonian formalism from a given hyperregular Lagrangian. Essential for this derivation were, so called, Legendre Transformations. Legendre Transformations are changes of variables associated with this Lagrangian.

Assume that a system is initially described by the position-velocity variables (q_i, v_i), $i = 1,...,N$, and its equations of motion are obtained from a Lagrangian $L(q,v)$ via its Euler-Lagrange equations (2.3), given in Chapter II. Then there exists a different set of variables $(\tilde{q}_i, p_i)$, $i = 1,...,N$, that is especially convenient. They are called "canonical variables." They are defined by:

$$\tilde{q}_i = q_i,$$
$$p_i = \frac{\partial L}{\partial v_i}, \qquad i = 1,...,N. \tag{H.1}$$

Comparing (H.1) and (3.1), we should observe that in (H.1) we introduced a different symbol for the "new" spatial variable on the left side of the top line. This way we avoid the situation where the "new" and "old" spatial variables are denoted by the same symbols. They are still equal numerically, because of the first line of (H.1), but now we have two different symbols. This makes further calculations less confusing. Still, (H.1) is the same Legendre Transformation as the one we considered in Chapter III.

A reader may ask why we even allow this confusion to be created in the first place and why we would not use different symbols all the way in this book. The answer is, we do not want to end up too far away from the traditional way these symbols are used. We see value in having our results look like they do in more traditional textbooks. Also, we see a value in having the intuition about the "position" described by one symbol, "q," no matter if it is followed by the velocity "v" or the canonical momentum "p."

Therefore, we decided not to use the symbol "$\tilde{q}$" in the entirety of our text. We will use it only here, in Appendix H, and only in the calculations that lead to the partial derivatives of the Hamiltonian.

As in Chapter III, Volume I, we assume that the Lagrangian is such that the formulas (H.1) are a change of variables. This means that we assume the formulas (H.1) to be invertible (at least locally) since invertibility is expected when we change variables. So, the equations (H.1) should allow an inverse which we will write as:

$$
\begin{aligned}
q_i &= q_i(\tilde{q}, p), \\
v_i &= v_i(\tilde{q}, p), \qquad i = 1, \ldots, N,
\end{aligned}
\tag{H.2}
$$

or more specifically:

$$
\begin{aligned}
q_i &= \tilde{q}_i, \\
v_i &= v_i(\tilde{q}, p), \qquad i = 1, \ldots, N.
\end{aligned}
\tag{H.3}
$$

Using the Lagrangian that gave us (H.1), (H.2), and (H.3), we define the Hamiltonian as:

$$
H(\tilde{q}, p) = \sum_{i=1}^{N} p_i \cdot v_i(\tilde{q}, p) - L(q, v) \Bigg|_{\substack{q_i = q_i(\tilde{q}, p) \\ v_i = v_i(\tilde{q}, p).}}
\tag{H.4}
$$

We can also write (H.4) as

$$
H(\tilde{q}, p) = \sum_{i=1}^{N} p_i \cdot v_i(\tilde{q}, p) - L(q(\tilde{q}, p), v(\tilde{q}, p)).
\tag{H.5}
$$

Since (H.1) and (H.2) represent a typical change of variables, we also get typical expressions for the partial derivatives. Using the regular chain rule for any function $F(q(\tilde{q}, p), v(\tilde{q}, p))$, we obtain:

$$\frac{\partial F(q(\tilde{q},p),v(\tilde{q},p))}{\partial \tilde{q}_n} = \sum_{i=1}^{N} \frac{\partial q_i(\tilde{q},p)}{\partial \tilde{q}_n} \cdot \frac{\partial F(q,v)}{\partial q_i} + \sum_{i=1}^{N} \frac{\partial v_i(\tilde{q},p)}{\partial \tilde{q}_n} \cdot \frac{\partial F(q,v)}{\partial v_i},$$

$$\frac{\partial F(q(\tilde{q},p),v(\tilde{q},p))}{\partial p_n} = \sum_{i=1}^{N} \frac{\partial q_i(\tilde{q},p)}{\partial p_n} \cdot \frac{\partial F(q,v)}{\partial q_i} + \sum_{i=1}^{N} \frac{\partial v_i(\tilde{q},p)}{\partial p_n} \cdot \frac{\partial F(q,v)}{\partial v_i}, \qquad i=1,\ldots,N.$$

(H.6)

Using the chain rule (H.6) to calculate the partial derivatives of the Hamiltonian (H.5), we get:

$$\frac{\partial H(\tilde{q},p)}{\partial \tilde{q}_n} = \frac{\partial}{\partial \tilde{q}_n}\left[\sum_{i=1}^{N} p_i \cdot v_i(\tilde{q},p)\right] - \frac{\partial L(q(\tilde{q},p),v(\tilde{q},p))}{\partial \tilde{q}_n} =$$

$$= \sum_{i=1}^{N} \frac{\partial p_i}{\partial \tilde{q}_n} \cdot v_i(\tilde{q},p) + \sum_{i=1}^{N} p_i \cdot \frac{\partial v_i(\tilde{q},p)}{\partial \tilde{q}_n} - \frac{\partial L(q(\tilde{q},p),v(\tilde{q},p))}{\partial \tilde{q}_n} =$$

$$= \sum_{i=1}^{N} 0 \cdot v_i(\tilde{q},p) + \sum_{i=1}^{N} p_i \cdot \frac{\partial v_i(\tilde{q},p)}{\partial \tilde{q}_n} - \frac{\partial L(q(\tilde{q},p),v(\tilde{q},p))}{\partial \tilde{q}_n} =$$

$$= \sum_{i=1}^{N} p_i \cdot \frac{\partial v_i(\tilde{q},p)}{\partial \tilde{q}_n} - \frac{\partial L(q(\tilde{q},p),v(\tilde{q},p))}{\partial \tilde{q}_n} =$$

(H.7)

using the chain rule:

$$= \sum_{i=1}^{N} p_i \cdot \frac{\partial v_i(\tilde{q},p)}{\partial \tilde{q}_n} - \sum_{i=1}^{N} \frac{\partial q_i(\tilde{q},p)}{\partial \tilde{q}_n} \cdot \frac{\partial L(q,v)}{\partial q_i} - \sum_{i=1}^{N} \frac{\partial v_i(\tilde{q},p)}{\partial \tilde{q}_n} \cdot \frac{\partial L(q,v)}{\partial v_i} =$$

(H.8)

using the specific form of (H.1), we can calculate some derivatives, getting:

$$= \sum_{i=1}^{N} p_i \cdot \frac{\partial v_i(\tilde{q},p)}{\partial \tilde{q}_n} - \sum_{i=1}^{N} \delta_{ni} \cdot \frac{\partial L(q,v)}{\partial q_i} - \sum_{i=1}^{N} \frac{\partial v_i(\tilde{q},p)}{\partial \tilde{q}_n} \cdot p_i =$$

(H.9)

where the symbol δ_{ni} is the Kronecker delta, so we get:

$$= \sum_{i=1}^{N} p_i \cdot \frac{\partial v_i(\tilde{q},p)}{\partial \tilde{q}_n} - \frac{\partial L(q,v)}{\partial q_n} - \sum_{i=1}^{N} p_i \cdot \frac{\partial v_i(\tilde{q},p)}{\partial \tilde{q}_n} =$$

(H.10)

cancelling the first and last terms gives:

$$= -\frac{\partial L(q,v)}{\partial q_n}, \qquad n = 1,\ldots,N. \tag{H.11}$$

Concluding the calculations (H.7) to (H.11), we get:

$$\frac{\partial H(\tilde{q},p)}{\partial \tilde{q}_n} = -\frac{\partial L(q,v)}{\partial q_n}, \qquad n = 1,\ldots,N. \tag{H.12}$$

The remaining partial derivatives of the Hamiltonian (H.5) can be calculated as:

$$\begin{aligned}\frac{\partial H(\tilde{q},p)}{\partial p_n} &= \frac{\partial}{\partial p_n}\left[\sum_{i=1}^{N} p_i \cdot v_i(\tilde{q},p)\right] - \frac{\partial L(q(\tilde{q},p),v(\tilde{q},p))}{\partial p_n} = \\ &= \sum_{i=1}^{N} \frac{\partial p_i}{\partial p_n} \cdot v_i(\tilde{q},p) + \sum_{i=1}^{N} p_i \cdot \frac{\partial v_i(\tilde{q},p)}{\partial p_n} - \frac{\partial L(q(\tilde{q},p),v(\tilde{q},p))}{\partial p_n} = \\ &= v_n(\tilde{q},p) + \sum_{i=1}^{N} p_i \cdot \frac{\partial v_i(\tilde{q},p)}{\partial p_n} - \frac{\partial L(q(\tilde{q},p),v(\tilde{q},p))}{\partial p_n} =\end{aligned} \tag{H.13}$$

using the chain rule:

$$\begin{aligned}&= v_n(\tilde{q},p) + \sum_{i=1}^{N} p_i \cdot \frac{\partial v_i(\tilde{q},p)}{\partial p_n} - \sum_{i=1}^{N} \frac{\partial q_i(\tilde{q},p)}{\partial p_n} \cdot \frac{\partial L(q,v)}{\partial q_i} - \sum_{i=1}^{N} \frac{\partial v_i(\tilde{q},p)}{\partial p_n} \cdot \frac{\partial L(q,v)}{\partial v_i} = \\ &= v_n(\tilde{q},p) + \sum_{i=1}^{N} p_i \cdot \frac{\partial v_i(\tilde{q},p)}{\partial p_n} - \sum_{i=1}^{N} 0 \cdot \frac{\partial L(q,v)}{\partial q_i} - \sum_{i=1}^{N} \frac{\partial v_i(\tilde{q},p)}{\partial p_n} \cdot \frac{\partial L(q,v)}{\partial v_i} = \\ &= v_n(\tilde{q},p) + \sum_{i=1}^{N} p_i \cdot \frac{\partial v_i(\tilde{q},p)}{\partial p_n} - \sum_{i=1}^{N} \frac{\partial v_i(\tilde{q},p)}{\partial p_n} \cdot \frac{\partial L(q,v)}{\partial v_i} =\end{aligned} \tag{H.14}$$

using (H.1) in the last term:

$$= v_n(\tilde{q},p) + \sum_{i=1}^{N} p_i \cdot \frac{\partial v_i(\tilde{q},p)}{\partial p_n} - \sum_{i=1}^{N} \frac{\partial v_i(\tilde{q},p)}{\partial p_n} \cdot p_i = \tag{H.15}$$

canceling the last two terms:

$$= v_n(\tilde{q}, p), \qquad n = 1, \ldots, N.$$

(H.17)

Concluding the calculations (H.13) to (H.17), we get:

$$\frac{\partial H(\tilde{q}, p)}{\partial p_n} = v_n(\tilde{q}, p), \qquad n = 1, \ldots, N. \tag{H.18}$$

We may observe that the results (H.12) and (H.18) for the partial derivatives of the Hamiltonian are identical to the results (3.10) and (3.13) from Chapter III, Volume I. However, here they are expressed in the variables described in (H.1), instead of the more traditional variables used in (3.1).

SUGGESTED READING

P.A.M. Dirac, "*Lectures on Quantum Mechanics*", (Belfer Graduate School, Yeshiva University, New York, 1964; reprinted by Dover Publications, 2001)

H. Goldstein, C. Poole, J. Safko "*Classical Mechanics*", 3rd edition, (Pearson 2001)

E.C.G. Sudarshan, N. Mukunda, "*Classical Dynamics: A Modern Perspective*", (World Scientific, 2015)

K. Sundermeyer, "*Constrained Dynamics*", (Springer-Verlag, 1982)

About the Author

Piotr Hebda is a professor of mathematics at the University of North Georgia, USA. He earned a master's degree in physics at the University of Wroclaw, Poland, and a master's degree in mathematics and a PhD in physics at the University of Georgia, USA.

His research interests are in classical mechanics of material points, with concentration on the Dirac's theory of constrained systems, and on using this theory for designing Lagrange and Hamiltonian formalisms for variety of mechanical systems.

www.ingramcontent.com/pod-product-compliance
Lightning Source LLC
Chambersburg PA
CBHW082202220526
45470CB00010B/3021